MEMOIRS
of the
American Mathematical Society

ISSN 0065-9266 (print)
ISSN 1947-6221 (online)

Number 1318

Local Well-Posedness and Break-Down Criterion of the Incompressible Euler Equations with Free Boundary

Chao Wang
Zhifei Zhang
Weiren Zhao
Yunrui Zheng

March 2021 • Volume 270 • Number 1318 (second of 7 numbers)

AMERICAN MATHEMATICAL SOCIETY

Library of Congress Cataloging-in-Publication Data

Cataloging-in-Publication Data has been applied for by the AMS.
See http://www.loc.gov/publish/cip/.
DOI: https://doi.org/10.1090/memo/1318

Memoirs of the American Mathematical Society

This journal is devoted entirely to research in pure and applied mathematics.

Subscription information. Beginning with the January 2010 issue, *Memoirs* is accessible from `www.ams.org/journals`. The 2021 subscription begins with volume 269 and consists of six mailings, each containing one or more numbers. Subscription prices for 2021 are as follows: for paper delivery, US$1085 list, US$868 institutional member; for electronic delivery, US$955 list, US$764 institutional member. Upon request, subscribers to paper delivery of this journal are also entitled to receive electronic delivery. If ordering the paper version, add US$20 for delivery within the United States; US$80 for outside the United States. Subscription renewals are subject to late fees. See `www.ams.org/help-faq` for more journal subscription information. Each number may be ordered separately; *please specify number* when ordering an individual number.

Back number information. For back issues see `www.ams.org/backvols`.

Subscriptions and orders should be addressed to the American Mathematical Society, P. O. Box 845904, Boston, MA 02284-5904 USA. *All orders must be accompanied by payment.* Other correspondence should be addressed to 201 Charles Street, Providence, RI 02904-2213 USA.

Memoirs of the American Mathematical Society (ISSN 0065-9266 (print); 1947-6221 (online)) is published bimonthly (each volume consisting usually of more than one number) by the American Mathematical Society at 201 Charles Street, Providence, RI 02904-2213 USA. Periodicals postage paid at Providence, RI. Postmaster: Send address changes to Memoirs, American Mathematical Society, 201 Charles Street, Providence, RI 02904-2213 USA.

This publication is indexed in *Mathematical Reviews®*, *Zentralblatt MATH*, *Science Citation Index®*, *Science Citation Index™-Expanded*, *ISI Alerting Services℠*, *SciSearch®*, *Research Alert®*, *CompuMath Citation Index®*, *Current Contents®/Physical, Chemical & Earth Sciences*.
This publication is archived in *Portico* and *CLOCKSS*.
Printed in the United States of America.

∞ The paper used in this book is acid-free and falls within the guidelines established to ensure permanence and durability.
Visit the AMS home page at `https://www.ams.org/`

10 9 8 7 6 5 4 3 2 1 26 25 24 23 22 21

Contents

Abstract

In this paper, we prove the local well-posedness of the free boundary problem for the incompressible Euler equations in low regularity Sobolev spaces, in which the velocity is a Lipschitz function and the free surface belongs to $C^{\frac{3}{2}+\varepsilon}$. Moreover, we also present a Beale-Kato-Majda type break-down criterion of smooth solution in terms of the mean curvature of the free surface, the gradient of the velocity and Taylor sign condition.

Received by the editor July 14, 2015, and, in revised form, November 12, 2017.
Article electronically published on May 21, 2021.
DOI: https://doi.org/10.1090/memo/1318

CHAPTER 1

Introduction

1.1. Presentation of the problem

In this paper, we consider the motion of an ideal incompressible gravity fluid in a domain with free boundary of finite depth

$$\{(t,x,y) \in [0,T] \times \mathbf{R}^d \times \mathbf{R} : (x,y) \in \Omega_t\},$$

where Ω_t is the fluid domain at time t located by

$$\Omega_t = \{(x,y) \in \mathbf{R}^d \times \mathbf{R} : b(x) < y < \eta(t,x)\}.$$

The motion of the fluid is described by the incompressible Euler equation

$$\partial_t v + v \cdot \nabla_{x,y} v = -g e_{d+1} - \nabla_{x,y} P \qquad \text{in} \quad \Omega_t,\ t \geq 0, \tag{1.1}$$

where $-ge_{d+1} = -g(0, \cdots, 0, 1)$ denotes the acceleration of gravity, $v = (v^1, ..., v^{d+1})$ denotes the velocity field, and P denotes the pressure. The incompressibility of the fluid is expressed by

$$\operatorname{div} v = 0 \qquad \text{in} \quad \Omega_t,\ t \geq 0. \tag{1.2}$$

Assume that no fluid particles are transported across the surface. At the bottom, this is given by

$$v_n|_{y=b(X)} := \mathbf{n}_- \cdot v|_{y=b(X)} = 0 \qquad \text{for} \quad t \geq 0, x \in \mathbf{R}^d \tag{1.3}$$

where $\mathbf{n}_- := \frac{1}{\sqrt{1+|\nabla_X b|^2}}(\nabla_X b, -1)^T$ denotes the outward normal vector to bottom of Ω_t. At the free surface, the boundary condition is kinematic and given by

$$\partial_t \eta - \sqrt{1+|\nabla\eta|^2} v_n|_{y=\eta(t,x)} = 0 \quad \text{for} \quad t \geq 0, x \in \mathbf{R}^d, \tag{1.4}$$

where $v_n = \mathbf{n}_+ \cdot v|_{y=\eta(t,x)}$, with $\mathbf{n}_+ := \frac{1}{\sqrt{1+|\nabla\eta|^2}}(-\nabla\eta, 1)^T$ denoting the outward normal vector to the free surface Σ_t. In general, the pressure at the free surface is proportional to the mean curvature of the free surface, i.e.,

$$P|_{y=\eta(t,x)} = -\kappa \nabla \cdot \Big(\frac{\nabla\eta}{\sqrt{1+|\nabla\eta|^2}}\Big) \qquad \text{for} \quad t \geq 0, x \in \mathbf{R}^d, \tag{1.5}$$

where $\kappa \geq 0$ is the surface tension coefficient.

In this paper, we consider the case without surface tension. Furthermore, we assume that the bottom is flat(i.e., $b(x) = -1$) in order to simplify our presentation. We also take the gravity constant $g = 1$.

1.2. Some known results

Let us first review some known results concerning the water-wave equations without vorticity. In the case when the surface tension is neglected and the motion of free surface is a small perturbation of still water, one could check Nalimov [**29**], Yosihara [**39**] and Craig [**18**] for the well-posedness of 2-D water-wave equations. In general, the local well-posedness of the water-wave equations of infinite depth without surface tension is solved by Wu [**34, 35**], where she shows that the Taylor sign condition

$$-\frac{\partial p}{\partial \mathbf{n}}\Big|_{y=\eta(t,x)} \geq c_0 > 0 \tag{1.6}$$

always holds as long as the free surface is no self-intersection. In [**7, 8**], Ambrose and Masmoudi present a different proof. Lannes [**24**] first solves the water-wave equations of finite depth without surface tension in the framework of the Eulerian coordinates. Ming and Zhang [**28**] generalize Lannes's result to the case with surface tension. In a series of works [**1–3**], Alazard, Burq and Zuily use the tools of paradifferential operators and Strichartz estimates to prove the local well-posedness of the water-wave equations in low regularity Sobolev spaces.

For small initial data, Wu [**36**] first proved the almost global well-posedness of 2-D water-wave equations. Wu [**37**] and Germain, Masmoudi and Shatah [**20**] proved the global well-posedness of 3-D water-wave equations by using different method. Recently, Alarzard and Delort [**6**] and Ionescu and Pusateri [**26**] independently proved the global well-posedness of 2-D water-wave equations, see also [**22, 23**] for a new proof based on the holomorphic coordinates. On the other hand, Castro, Cordoba, Ferferman, Gancedo and Lopez-Fernandez [**12**] show that there exists smooth initial data for the water-waves equations such that the solution overturns in finite time. See [**13, 17**] for the formation of the splash singularity. Wu [**38**] also constructed a class of self-similar solution for the 2-D water-wave equations without the gravity.

Now, we review some well-posedness results for the rotational water-wave equations. Christodoulou and Lindblad [**15**] presented the a priori estimates of the incompressible Euler equations in a free domain diffeomorphic to a ball. Later, Lindblad [**27**] proved the local existence of smooth solution by using Nash-Moser iteration. Coutand and Shkoller [**16**] proved the local well-posedness of the incompressible Euler equations in both cases with surface tension and without surface tension by using the lagrangian coordinates and a subtle mollification procedure. Zhang and Zhang [**40**] solve the incompressible Euler equations without surface tension by using the framework of Clifford analysis introduced by Wu [**35**]. Shatah and Zeng [**30–32**] solve this problem by deriving the evolution equations of geometry quantities, especially the mean curvature.

In this memoir, we first prove the local well-posedness of the rotational water-wave problem in low regularity Sobolev spaces, and then present a break-down criterion to the obtained smooth solution in terms of physical quantity and geometrical quantity. This work was motivated by Craig and Wayne's question proposed in [**19**]: "**How do solutions break down?**"

There are several versions of this question, including " What is the lowest exponent of a Sobolev space H^s in which one can produce an existence theorem local in time?" Or one could ask "For which α is it true that, if one knows a priori *that* $\sup_{[-T,T]} \|(\eta,\psi)\|_{C^\alpha} < +\infty$ *and that* $(\eta_0,\psi_0) \in C^\infty$, *then the solution is fact* C^∞ *over the time interval* $[-T,T]$*?" ······ It would be more satisfying to say that the solution fails to exist because the curvature of the surface has diverged at some point, or a related geometrical and(or) physical statement.*

In the case without the vorticity and surface tension, this question was solved by Alazard, Burq and Zuily for the low regularity well-posedness [**3**], and by Wang and Zhang for the break-down criterion [**33**].

1.3. Main results

The first main result of this memoir is the local well-posedness of the water-wave equations in Sobolev spaces with low regularity, where the regularity of the initial velocity is consistent with the classical local well-posedness result in $\mathbf{R}^{d+1}$ proved by Kato and Ponce [**21**].

THEOREM 1.1. *Let* $d \geq 1$ *and* $s > \frac{d}{2}+1$. *Assume that the initial data* (η_0, v_0) *satisfies*

$$\eta_0 \in H^{s+\frac{1}{2}}(\mathbf{R}^d), \quad v_0 \in H^{s+\frac{1}{2}}(\Omega_0).$$

Furthermore, assume that there exist two positive constants $c_0 > 0$ *and* $h_0 > 0$ *such that*

$$-(\partial_y P)(0,x,\eta_0(x)) \geq c_0 \quad \text{for } x \in \mathbf{R}^d, \tag{1.7}$$

$$1+\eta_0(x) \geq h_0 \quad \text{for } x \in \mathbf{R}^d. \tag{1.8}$$

Then there exists $T > 0$ *such that the system* (1.1)*–*(1.5) *with the initial data* (η_0, v_0) *has a unique solution* (η, v) *satisfying*

$$\eta \in C\big([0,T]; H^{s+\frac{1}{2}}(\mathbf{R}^d)\big), \quad \widetilde{v} \in C\big([0,T]; H^{s+\frac{1}{2}}(\overline{\mathcal{S}})\big),$$

where $\widetilde{v}(t,x,z) = v(t,x,\rho_\delta(t,x,z))$ *for* $(x,z) \in \overline{\mathcal{S}} = \mathbf{R}^d \times I$ *with* $I = [-1,0]$*(see section 4.2 for the definition of* ρ_δ*).*

REMARK 1.2. The regularity of the initial velocity should be optimal by the recent strong ill-posedness result in the whole space proved by Bourgain and Li [**11**]. The regularity of the initial free surface could be further lowered by using the Strichartz type estimates, see [**5**] for the irrotational case.

In a seminal paper [**10**], Beale, Kato and Majda showed that if v is a smooth solution of the incompressible Euler equations in $[0,T) \times \mathbf{R}^3$ and satisfies

$$\int_0^T \|\nabla \times v(t)\|_{L^\infty(\mathbf{R}^3)} dt < +\infty,$$

then the solution can be extended after $t = T$. The second main result of this paper is a Beale-Kato-Majda type blow-up criterion for the free boundary problem of the incompressible Euler equations.

THEOREM 1.3. *Let* $s > \frac{d}{2}+1$ *so that* $s-\frac{1}{2}$ *is an integer, and* (η, v, P) *be the solution of the system* (1.1)*–*(1.5) *in* $[0,T]$ *obtained in Theorem* 1.1. *If the solution*

(η, v, P) satisfies

$$M(T) \stackrel{def}{=} \sup_{t\in[0,T]} \big(\|H(t)\|_{L^p\cap L^2} + \|v(t)\|_{W^{1,\infty}(\Omega_t)}\big) < +\infty,$$

$$\inf_{(t,x,y)\in[0,T]\times\Sigma_t} -\frac{\partial P}{\partial \mathbf{n}}(t,x,y) \geq c_0,$$

$$1+\eta(t,x) \geq h_0 \quad \text{for } x \in \mathbf{R}^d,$$

for some $p > 2d$ and $c_0 > 0, h_0 > 0$, then it holds that

$$\sup_{t\in[0,T]} E_s(t) \leq C\big(E_s(0), M(T), T, c_0, h_0\big),$$

Especially, the solution (η, v) can be extended after $t = T$. Here $H(t,x)$ is the mean curvature of the free surface and

$$E_s(t) \stackrel{def}{=} \|\eta(t)\|_{H^{s+\frac{1}{2}}} + \|v(t)\|_{H^{s+\frac{1}{2}}(\Omega(t))}.$$

1.4. Main ideas

We denote by (V, B) the horizontal and vertical traces of the velocity on the free surface, i.e.,

$$V \triangleq (v^1, \cdots, v^d)|_{y=\eta}, \quad B \triangleq v^{d+1}|_{y=\eta}.$$

Introduce a good unknown $U = V + T_\zeta B$, where $\zeta = \nabla\eta$ and T_ζ is Bony's paraproduct. We can derive the following evolution equation for U:

$$D_t^2 U + T_{a\lambda} U = f + f_\omega. \tag{1.9}$$

Here $D_t = \partial_t + T_V \cdot \nabla$, $T_{a\lambda}$ is an elliptic paradifferential operator of order one, and f_ω is the nonlinear term induced by the vorticity.

Compared with the irrotational case, a main difficulty is that f_ω lose one half derivative. More precisely, $f \in H^{s-\frac{1}{2}}$ but $f_\omega \in H^{s-1}$. Our key observation is that $D_t f_\omega$ has the same regularity as f_ω, and $\|f_\omega\|_{H^{s-1}}$ can be controlled by the lower order energy. By using the following trick

$$\begin{aligned}\big\langle \langle D\rangle^{s-\frac{1}{2}} f_\omega, \langle D\rangle^{s-\frac{1}{2}} D_t U\big\rangle =& \frac{d}{dt}\big\langle \langle D\rangle^{s-1} f_\omega, \langle D\rangle^{s} U\big\rangle - \big\langle \langle D\rangle^{s-1} D_t f_\omega, \langle D\rangle^{s} U\big\rangle \\ &+ \text{Lower order terms},\end{aligned}$$

we can obtain an energy inequality of the form

$$E(t) \leq \big\langle \langle D\rangle^{s-1} f_\omega, \langle D\rangle^{s} U\big\rangle + E(0) + \int_0^t \mathcal{P}(E(t'))dt', \tag{1.10}$$

where $\mathcal{P}$ is an increasing function. The term $\big\langle \langle D\rangle^{s-1} f_\omega, \langle D\rangle^{s} U\big\rangle$ can be controlled by

$$\|f_\omega\|_{H^{s-1}}\|U\|_{H^s} \leq \mathcal{P}(E_l(t))E(t)^{\frac{1}{2}} \leq \mathcal{P}(E_l(t)) + \frac{1}{2}E(t).$$

Here $E_l(t)$ is a lower order energy, which satisfies

$$E_l(t) \leq E_l(0) + \int_0^t \mathcal{P}(E(t'))dt'.$$

This together with (1.10) gives a close estimate for $E(t)$.

Compared with the work [**3**], a new technical ingredient is that we introduce Chemin-Lerner type Besov spaces in the elliptic estimates such that we can obtain the maximal Hölder regularity estimates, which play an important role in the proof of break-down criterion. Otherwise, if we just follow the framework of [**3**], it is possible to establish a similar break-down criterion, where $\|v(t)\|_{W^{1,\infty}(\Omega_t)}$ is replaced by $\|v(t)\|_{C^{1,\alpha}(\Omega_t)}$ for some $\alpha > 0$.

For the free boundary problem, it is highly nontrivial to obtain the existence of the solution from a priori estimates. The main reason is that many special structures of the system are used in the process of a priori estimates, however it is usually difficult to keep these structures for the approximate system.

To construct the approximate system, an immediate idea is that we use the equation (1.9) of U to construct the approximate system for the unknowns defined on the free surface. However, we find that it is difficult to show that the limit system is equivalent to the original Euler system. Instead, we still use the first order system to construct the iteration scheme. In order to keep as more structures of the system as possible, a key idea is that more unknowns and new equations are introduced such that the structures are integrated into the new equations. The construction of the iteration scheme are very tricky, where we also used another important observation: the maximal regularity of the free surface is not needed for the estimate of the vorticity and the velocity, and it is just used in the estimate of the pressure.

CHAPTER 2

Tools of paradifferential operators

In this chapter, we introduce some basic results about the paradifferential operators from [**25**](see also [**3**]).

2.1. Paradifferential operators

Let us first introduce the definition of the symbol with limited spatial smoothness. We denote by $W^{k,\infty}(\mathbf{R}^d)$ the usual Sobolev spaces for $k \in \mathbf{N}$, and the Hölder space C^s with exponent s for $s \in (0,1)$.

DEFINITION 2.1. Given $\mu \in [0,1]$ and $m \in \mathbf{R}$, we denote by $\Gamma^m_\mu(\mathbf{R}^d)$ the space of locally bounded functions $a(x,\xi)$ on $\mathbf{R}^d \times \mathbf{R}^d \backslash \{0\}$, which are C^∞ with respect to ξ for $\xi \neq 0$ and such that, for all $\alpha \in \mathbf{N}^d$ and all $\xi \neq 0$, the function $x \to \partial^\alpha_\xi a(x,\xi)$ belongs to $W^{\mu,\infty}$ and there exists a constant C_α such that

$$\|\partial^\alpha_\xi a(\cdot,\xi)\|_{W^{\mu,\infty}} \leq C_\alpha (1+|\xi|)^{m-|\alpha|} \quad \text{for any} \quad |\xi| \geq \frac{1}{2}.$$

The semi-norm of the symbol is defined by

$$M^m_\mu(a) \stackrel{\text{def}}{=} \sup_{|\alpha| \leq 3d/2+1+\mu} \sup_{|\xi|\geq 1/2} \|(1+|\xi|)^{|\alpha|-m} \partial^\alpha_\xi a(\cdot,\xi)\|_{W^{\mu,\infty}}.$$

Especially, if a is a function independent of ξ, then

$$M^m_\mu(a) = \|a\|_{W^{\mu,\infty}}.$$

Given a symbol a, the paradifferential operator T_a is defined by

$$\widehat{T_a u}(\xi) \stackrel{\text{def}}{=} (2\pi)^{-d} \int \chi(\xi-\eta,\eta)\widehat{a}(\xi-\eta,\eta)\psi(\eta)\widehat{u}(\eta)d\eta, \tag{2.1}$$

where $\widehat{a}(\theta,\xi)$ is the Fourier transform of a with respect to the first variable; the $\chi(\theta,\xi) \in C^\infty(\mathbf{R}^d \times \mathbf{R}^d)$ is an admissible cut-off function: there exists $\varepsilon_1, \varepsilon_2$ such that $0 < \varepsilon_1 < \varepsilon_2$ and

$$\chi(\theta,\eta) = 1 \quad \text{for} \quad |\theta| \leq \varepsilon_1|\eta|, \quad \chi(\theta,\eta) = 0 \quad \text{for} \quad |\theta| \geq \varepsilon_2|\eta|,$$

and such that for any $(\theta,\eta) \in \mathbf{R}^d \times \mathbf{R}^d$,

$$|\partial^\alpha_\theta \partial^\beta_\eta \chi(\theta,\eta)| \leq C_{\alpha,\beta}(1+|\eta|)^{-|\alpha|-|\beta|}.$$

The cut-off function $\psi(\eta) \in C^\infty(\mathbf{R}^d)$ satisfies

$$\psi(\eta) = 0 \quad \text{for} \quad |\eta| \leq 1, \quad \psi(\eta) = 1 \quad \text{for} \quad |\eta| \geq 2.$$

Here we will take the admissible cut-off function $\chi(\theta,\eta)$ as follows

$$\chi(\theta,\eta) = \sum_{k=0}^\infty \zeta_{k-3}(\theta)\varphi_k(\eta),$$

where $\zeta(\theta) = 1$ for $|\theta| \le 1.1$ and $\zeta(\theta) = 0$ for $|\theta| \ge 1.9$; and

$$\zeta_k(\theta) = \zeta(2^{-k}\theta) \quad \text{for} \quad k \in \mathbf{Z},$$
$$\varphi_0 = \zeta, \quad \varphi_k = \zeta_k - \zeta_{k-1} \quad \text{for} \quad k \ge 1.$$

We also introduce the Littlewood-Paley operators Δ_k, S_k defined by

$$\Delta_k u = \mathcal{F}^{-1}\big(\varphi_k(\xi)\widehat{u}(\xi)\big) \quad \text{for } k \ge 0, \quad \Delta_k u = 0 \quad \text{for } k < 0,$$
$$S_k u = \sum_{\ell \le k} \Delta_\ell u \quad \text{for } k \in \mathbf{Z}.$$

In the case when the function a depends only on the first variable x in $T_a u$, we take $\psi = 1$. Then $T_a u$ is just the usual Bony's paraproduct defined by

$$T_a u = \sum_k S_{k-3} a \Delta_k u. \tag{2.2}$$

Furthermore, we have Bony's decomposition:

$$au = T_a u + T_u a + R(u, a), \tag{2.3}$$

where the remainder term $R(u, a)$ is defined by

$$R(u, a) = \sum_{|k-\ell| \le 2} \Delta_k a \Delta_\ell u. \tag{2.4}$$

The following Berstein's inequality will be repeatedly used.

LEMMA 2.2. *Let $1 \le p \le q \le \infty, \alpha \in \mathbf{N}^d$. Then it holds that*

$$\|\partial^\alpha S_k u\|_{L^q} \le C 2^{kd(\frac{1}{p} - \frac{1}{q} + |\alpha|)} \|S_k u\|_{L^p} \quad \textit{for } k \in \mathbf{N},$$
$$\|\Delta_k u\|_{L^q} \le C 2^{kd(\frac{1}{p} - \frac{1}{q} - |\alpha|)} \sup_{|\beta| = |\alpha|} \|\partial^\beta \Delta_k u\|_{L^p} \quad \textit{for } k \ge 1.$$

2.2. Functional spaces

We introduce some functional spaces, which will be used throughout this paper.

DEFINITION 2.3. Let $s \in \mathbf{R}$ and $p, q \in [1, \infty]$. The inhomogeneous Besov space $B^s_{p,q}(\mathbf{R}^d)$ is defined by

$$B^s_{p,q}(\mathbf{R}^d) \stackrel{\text{def}}{=} \Big\{ f \in \mathcal{S}'(\mathbf{R}^d) : \|f\|_{B^s_{p,q}} \triangleq \big(\sum_j 2^{jsq} \|\Delta_j f\|^q_{L^p}\big)^{\frac{1}{q}} < \infty \Big\}.$$

In the case of $p = q = 2$, $B^s_{p,q}(\mathbf{R}^d)$ is just the usual Sobolev space $H^s(\mathbf{R}^d)$; In the case of $p = q = \infty$, $B^s_{p,q}(\mathbf{R}^d)$ is the Zygmund space $C^s(\mathbf{R}^d)$.

Let $\mathcal{S} = \mathbf{R}^d \times I$ with $I \subset \mathbf{R}$ an interval. We introduce the Sobolev space $H^s(\mathcal{S})$ on $\mathcal{S}$. When s is an integer, $H^s(\mathcal{S})$ is just the usual Sobolev space. In general case, let $k = [s]$ and $\sigma = s - k \in (0, 1)$. The norm of $H^s(\mathcal{S})$ is defined by

$$\|u\|_{H^s(\mathcal{S})} \stackrel{\text{def}}{=} \sum_{\ell \le k} \|\nabla^\ell_{x,z} u\|_{L^2_z(I; H^\sigma)} + \Big(\int_I \int_I \int_{\mathbf{R}^d} \frac{|\nabla^k_{x,z} u(x,z) - \nabla^k_{x,z} u(x,z')|^2}{|z - z'|^{1+2\sigma}} dx dz dz' \Big)^{\frac{1}{2}}.$$

In order to obtain the optimal elliptic regularity, let us introduce Chemin-Lerner type Besov space $\widetilde{L}^q_z(I; B^s_{p,r}(\mathbf{R}^d))$, whose norm is defined by

$$\|f\|_{\widetilde{L}^q_z(I; B^s_{p,r})} \stackrel{\text{def}}{=} \Big(\sum_k 2^{ksr} \|\Delta_k f\|^r_{L^q_z(I; L^p)} \Big)^{\frac{1}{r}}.$$

In the case of $p=r=\infty$, we denote it by $\widetilde{L}^q_z(I;C^s(\mathbf{R}^d))$; In the case of $p=q=r=2$, we have $\widetilde{L}^q_z(I;B^s_{p,r}(\mathbf{R}^d))\equiv L^2_z(I;H^s(\mathbf{R}^d))$; When $q=\infty, p=r=2$, we denote it by $\widetilde{L}^\infty_z(I;H^s(\mathbf{R}^d))$. In this case, there holds

$$\|f\|_{L^\infty_z(I;H^s)}\le \|f\|_{\widetilde{L}^\infty_z(I;H^s)}.$$

This kind of space was firstly introduced by Chemin and Lerner [**14**] to study the incompressible Navier-Stokes equations.

The following characterization of Sobolev space is very useful.

LEMMA 2.4. *Let $s\in\mathbf{R}$ and $c>0$. Suppose that $\{u_k\}_{k\in\mathbf{N}}$ is a sequence of functions in $L^2(\mathbf{R}^d)$ such that (1) u_0 is spectrally supported in a ball $\{|\xi|\le c^{-1}\}$ and u_k for $k>0$ is spectrally supported in an annulus $\{c2^k\le|\xi|\le c^{-1}2^k\}$; (2) $\{2^{ks}\|u_k\|_{L^2}\}_{k\in\mathbf{N}}\in\ell^2$. Then $u=\sum_k u_k\in H^s(\mathbf{R}^d)$ and*

$$\|u\|_{H^s}\le C\Big(\sum_k 2^{2ks}\|u_k\|^2_{L^2}\Big)^{\frac12}.$$

In addition, for $s>0$, it is sufficient to assume that u_k is spectrally supported in a ball $\{|\xi|\le c^{-1}2^k\}$.

We also introduce the anisotropic Sobolev space $H^{s,\sigma}(\mathbf{R}^{d+1})$, whose norm is given by

$$\|u\|_{H^{s,\sigma}}\stackrel{\text{def}}{=}\|\langle D_x\rangle^\sigma u\|_{H^s(\mathbf{R}^{d+1})},\quad x\in\mathbf{R}^d.$$

We also have a similar characterization.

LEMMA 2.5. *For any $s,\sigma\in\mathbf{R}$, we have*

$$\|u\|^2_{H^{s,\sigma}}\sim\sum_{\ell,j}2^{2\ell s}2^{2j\sigma}\|\Delta_\ell\Delta^h_j u\|^2_{L^2}.$$

Here Δ^h_j is the Littlewood-Paley operator in the x direction. If $s,\sigma>0$, and the sequence $\{u_{\ell,j}\}_{\ell,j\in\mathbf{N}}$ is spectrally supported in $\{|\xi|\le c^{-1}2^\ell\}\cap\{|\xi_h|\le c^{-1}2^j\}$ for some $c>0$ and satisfies $\|u_{\ell,j}\|_{L^2}\le c_{\ell,j}2^{-\ell s}2^{-j\sigma}$ with $\{c_{\ell,j}\}_{\ell,j\in\mathbf{N}}\in\ell^2$, then $u=\sum_{\ell,j}u_{\ell,j}\in H^{s,\sigma}(\mathbf{R}^{d+1})$ satisfies

$$\|u\|_{H^{s,\sigma}}\le C\Big(\sum_{\ell,j}2^{2\ell s}2^{2j\sigma}\|u_{\ell,j}\|^2_{L^2}\Big)^{\frac12}.$$

2.3. Symbolic calculus

Let us recall the symbolic calculus and the boundedness in Sobolev space and Besov space of the paradifferential operators.

PROPOSITION 2.6. *Let $m,m'\in\mathbf{R}$.*

1. *If $a\in\Gamma^m_0(\mathbf{R}^d)$, then for any $s\in\mathbf{R}$,*

$$\|T_a\|_{H^s\to H^{s-m}}\le CM^m_0(a).$$

2. *If $a\in\Gamma^m_\rho(\mathbf{R}^d), b\in\Gamma^{m'}_\rho(\mathbf{R}^d)$ for $\rho>0$, then for any $s\in\mathbf{R}$,*

$$\|T_aT_b-T_{a\#b}\|_{H^s\to H^{s-m-m'+\rho}}\le CM^{m_1}_\rho(a)M^{m'}_0(b)+CM^{m_1}_0(a)M^{m'}_\rho(b),$$

where $a\#b=\sum_{|\alpha|<\rho}\partial^\alpha_\xi a(x,\xi)D^\alpha_x b(x,\xi), D_x=\frac{\partial_x}{i}$.

3. *If* $a \in \Gamma_\rho^m(\mathbf{R}^d)$ *for* $\rho \in (0,1]$, *then for any* $s \in \mathbf{R}$,

$$\|T_{a^*} - (T_a)^*\|_{H^s \to H^{s-m+\rho}} \le CM_\rho^m(a).$$

Here $(T_a)^*$ *is the adjoint operator of* T_a, *and* C *is a constant independent of* a, b.

PROPOSITION 2.7. [**33**] *Let* $m, m', s \in \mathbf{R}, q \in [1,\infty]$ *and* $\rho \in [0,1]$.

1. *If* $a \in \Gamma_0^m(\mathbf{R}^d)$, *then*

$$\|T_a\|_{B^s_{\infty,q} \to B^{s-m}_{\infty,q}} \le CM_0^m(a);$$

2. *If* $a \in \Gamma_\rho^m(\mathbf{R}^d), b \in \Gamma_\rho^{m'}(\mathbf{R}^d)$, *then*

$$\|T_aT_b - T_{ab}\|_{B^s_{\infty,q} \to B^{s-m-m'+\rho}_{\infty,q}} \le CM_\rho^m(a)M_0^{m'}(b) + CM_0^m(a)M_\rho^{m'}(b).$$

Here C *is a constant independent of* a, b.

REMARK 2.8. If the symbol $a(x,\xi)$ satisfies

$$M^m_{-\mu}(a) \triangleq \sup_{|\alpha| \le 3d/2+1} \sup_{|\xi| \ge 1/2} \|(1+|\xi|)^{|\alpha|-m} \partial_\xi^\alpha a(\cdot,\xi)\|_{C^{-\mu}} < \infty$$

for some $\mu > 0$, then T_a is bounded from $H^s(\mathbf{R}^d)$ to $H^{s-m-\mu}(\mathbf{R}^d)$ and $B^s_{\infty,q}(\mathbf{R}^d)$ to $B^{s-m-\mu}_{\infty,q}(\mathbf{R}^d)$ with the bound $M^m_{-\mu}(a)$.

COROLLARY 2.9. *Let* $s, m_1, m_2, m_3 \in \mathbf{R}$. *Suppose that* $a \in \Gamma_1^{m_1}, b \in \Gamma_2^{m_2}, c \in \Gamma_2^{m_3}$. *Then we have*

$$\big\| [T_a, [T_b, T_c]] \big\|_{H^s \to H^{s-m_1-m_2-m_3+2}} \le CM_1^{m_1}(a)M_2^{m_2}(b)M_2^{m_3}(c).$$

PROOF. It follows from Proposition 2.6 that

$$\|[T_b, T_c] - T_p + T_{p_1}\|_{H^s \to H^{s-m_2-m_3+2}} \le M_2^{m_2}(b)M_2^{m_3}(c).$$

where $p(x,\xi) = \sum_{|\alpha|=1} \partial_\xi^\alpha b(x,\xi) D_x^\alpha c(x,\xi)$ and $p_1(x,\xi) = \sum_{|\alpha|=1} \partial_\xi^\alpha c(x,\xi) D_x^\alpha b(x,\xi)$. Hence, it is sufficient to consider $[T_a, T_p]$ and $[T_a, T_{p_1}]$. Because of $p, p_1 \in \Gamma_1^{m_2+m_3-1}$, the corollary follows from Proposition 2.6. □

2.4. Tame estimates in Sobolev space

Let us first recall some classical tame estimates. One can refer to [**9**] for more general results.

LEMMA 2.10. *Let* $s \in \mathbf{R}$, *and* $p, q \in [1,\infty]$. *Then for any* $\sigma > 0$, *we have*

$$\|T_u v\|_{B^s_{p,q}} \le C \min\big(\|u\|_{C^{-\sigma}} \|v\|_{B^{s+\sigma}_{p,q}}, \|u\|_{L^\infty} \|v\|_{B^s_{p,q}}\big).$$

If $s_1 + s_2 > 0$, *then we have*

$$\|R(u,v)\|_{H^{s_1+s_2-\frac{d}{2}}} \le C\|u\|_{H^{s_1}} \|v\|_{H^{s_2}}.$$

If $s > 0$, *then for any* $\sigma \in \mathbf{R}$, *we have*

$$\|R(u,v)\|_{B^s_{p,q}} \le C\|u\|_{C^\sigma} \|v\|_{B^{s-\sigma}_{p,q}}.$$

PROOF. The first two inequalities are classical, see [**9**] for example. We prove the third inequality. Recall that

$$R(u,v) = \sum_{|k-\ell| \le 2} \Delta_k u \Delta_\ell v.$$

Hence, there exits some $N_0 \in \mathbf{N}$ so that

$$\Delta_j R(u,v) = \sum_{|k-\ell|\le 2; k,\ell\ge j-N_0} \Delta_j\big(\Delta_k u \Delta_\ell v\big).$$

It follows from Lemma 2.2 that

$$\begin{aligned}\|\Delta_j R(u,v)\|_{L^p} \le & \sum_{|k-\ell|\le 2; k,\ell\ge j-N_0} \|\Delta_k u\|_{L^\infty}\|\Delta_\ell v\|_{L^p}\\ \le & C\|u\|_{C^\sigma} \sum_{\ell\ge j-N_0} 2^{-\sigma\ell}\|\Delta_\ell v\|_{L^p}.\end{aligned}$$

This gives

$$2^{js}\|\Delta_j R(u,v)\|_{L^p} \le C\|u\|_{C^\sigma} \sum_{\ell\ge j-N_0} 2^{-s(\ell-j)}2^{\ell(s-\sigma)}\|\Delta_\ell v\|_{L^p},$$

which implies the second inequality by Young's inequality. □

A direct consequence of Lemma 2.10 is the following tame product estimate.

LEMMA 2.11. *Let $s \ge 0$. Then we have*

$$\begin{aligned}&\|fg\|_{H^s} \le C\big(\|f\|_{L^\infty}\|g\|_{H^s} + \|g\|_{L^\infty}\|f\|_{H^s}\big),\\ &\|f\nabla g\|_{H^s} \le C\big(\|f\|_{L^\infty}\|\nabla g\|_{H^s} + \|g\|_{L^\infty}\|f\|_{H^{s+1}}\big).\end{aligned}$$

LEMMA 2.12. [9] *Let $s > 0, p, q \in [1,\infty]$ and F be a smooth function with $F(0) = 0$. Then we have*

$$\|F(u)\|_{B^s_{p,q}} \le C(\|u\|_{L^\infty})\|u\|_{B^s_{p,q}}.$$

Especially, for $p = q = 2$, we have

$$\|F(u)\|_{H^s} \le C(\|u\|_{L^\infty})\|u\|_{H^s}.$$

Using an extension argument, we deduce from Lemma 2.11 and Lemma 2.12 that

LEMMA 2.13. *Let $\mathcal{S} = \mathbf{R}^d \times I$ with $I \subset \mathbf{R}$ an interval and $s \ge 0$. Then we have*

$$\begin{aligned}&\|uv\|_{H^s(\mathcal{S})} \le C\big(\|u\|_{L^\infty(\mathcal{S})}\|v\|_{H^s(\mathcal{S})} + \|v\|_{L^\infty(\mathcal{S})}\|u\|_{H^s(\mathcal{S})}\big),\\ &\|u\nabla v\|_{H^s(\mathcal{S})} \le C\big(\|u\|_{L^\infty(\mathcal{S})}\|\nabla v\|_{H^s(\mathcal{S})} + \|v\|_{L^\infty(\mathcal{S})}\|u\|_{H^{s+1}(\mathcal{S})}\big).\end{aligned}$$

Let F be a smooth function with $F(0) = 0$. Then we have

$$\|F(u)\|_{H^s(\mathcal{S})} \le C(\|u\|_{L^\infty(\mathcal{S})})\|u\|_{H^s(\mathcal{S})}.$$

Next we present a tame estimate in the anisotropic Sobolev space.

LEMMA 2.14. *Let $s > 0$ and $\sigma \in (0,1)$. Then it holds that for any $\epsilon > 0$,*

$$\begin{aligned}&\|T_u v\|_{H^{s,\sigma}} + \|R(u,v)\|_{H^{s,\sigma}}\\ &\le C\big(\|u\|_{L^\infty} + \|\langle D_x\rangle^{\sigma+\epsilon}u\|_{L^\infty_y(\mathbf{R};L^2)} + \|\langle D_x\rangle^{\frac{d}{2}}u\|_{L^\infty_y(\mathbf{R};L^2)}\big)\|v\|_{H^{s,\sigma}}.\end{aligned}$$

PROOF. Using Bony's decomposition (2.3), we write

$$T_u v = T(T^h + \overline{T}^h + R^h)(u,v).$$

where $T(u,v) = T_u v, \overline{T}^h(u,v) = T^h(v,u)$, and $T^h_u v$ denote the paraproduct in the x direction.

By using Lemma 2.2 and Lemma 2.5, it is easy to show that

$$\|(TT^h + TR^h)(u,v)\|_{H^{s,\sigma}} \le C\|u\|_{L^\infty}\|v\|_{H^{s,\sigma}}. \tag{2.5}$$

According to the definition of the paraproduct, we have

$$(T\overline{T}^h)(u,v) = \sum_{\ell,j} S_{\ell-3}\Delta_j^h u \Delta_\ell S_{j-3}^h v.$$

For $\sigma < \frac{d}{2}$, we get by Lemma 2.2 that

$$\begin{aligned}
\|S_{\ell-3}\Delta_j^h u\Delta_\ell S_{j-3}^h v\|_{L^2(\mathbf{R}^{d+1})} &\le \|S_{\ell-3}\Delta_j^h u\|_{L^\infty_y(\mathbf{R};L^2)}\|\Delta_\ell S_{j-3}^h v\|_{L^2_y(\mathbf{R};L^\infty)}\\
&\le C2^{j(\frac{d}{2}-\sigma)}\|\Delta_j^h u\|_{L^\infty_y(\mathbf{R};L^2)}\|\langle D_x\rangle^\sigma\Delta_\ell v\|_{L^2}\\
&\le C2^{-j\sigma}\|\langle D_x\rangle^{\frac{d}{2}}\Delta_j^h u\|_{L^\infty_y(\mathbf{R};L^2)}\|\langle D_x\rangle^\sigma\Delta_\ell v\|_{L^2}\\
&\le Cc_{\ell,j}2^{-\ell s}2^{-j\sigma}\|\langle D_x\rangle^{\frac{d}{2}}u\|_{L^\infty_y(\mathbf{R};L^2)}\|v\|_{H^{s,\sigma}},
\end{aligned}$$

and for $\sigma \ge \frac{d}{2}$ and any $\epsilon > 0$, we have

$$\begin{aligned}
\|S_{\ell-3}\Delta_j^h u\Delta_\ell S_{j-3}^h v\|_{L^2(\mathbf{R}^{d+1})} &\le \|S_{\ell-3}\Delta_j^h u\|_{L^\infty_y(\mathbf{R};L^2)}\|\Delta_\ell S_{j-3}^h v\|_{L^2_z(\mathbf{R};L^\infty)}\\
&\le C2^{j\epsilon}\|\Delta_j^h u\|_{L^\infty_y(\mathbf{R};L^2)}\|\langle D_x\rangle^{\frac{d}{2}-\epsilon}\Delta_\ell v\|_{L^2}\\
&\le Cc_{\ell,j}2^{-\ell s}2^{-j\sigma}\|\langle D_x\rangle^{\sigma+\epsilon}u\|_{L^\infty_y(\mathbf{R};L^2)}\|v\|_{H^{s,\sigma}},
\end{aligned}$$

where $\|\{c_{\ell,j}\}\|_{\ell^2} \le 1$. Then Lemma 2.5 ensures that

$$\|(T\overline{T}^h)(u,v)\|_{H^{s,\sigma}} \le C\big(\|\langle D_x\rangle^{\sigma+\epsilon}u\|_{L^\infty_y(\mathbf{R};L^2)} + \|\langle D_x\rangle^{\frac{d}{2}}u\|_{L^\infty_y(\mathbf{R};L^2)}\big)\|v\|_{H^{s,\sigma}},$$

which together with (2.5) gives the first part of the lemma. The proof of another part is similar. □

Next, we give the material derivative estimates for $R(u,v)$.

LEMMA 2.15. *Let $\overline{\partial}_t \triangleq \partial_t + V\cdot\nabla$. Let $s > 0$ and $\sigma_1 \in (0,1), \sigma_2 \in \mathbf{R}$ and $\sigma > 1-\sigma_1$. Then it holds that*

$$\begin{aligned}
\|\overline{\partial}_t R(u(t),v(t))\|_{H^s} &\le C\big(\|\overline{\partial}_t u\|_{C^{\sigma_1}}\|v\|_{H^{s-\sigma_1}} + \|\overline{\partial}_t v\|_{C^{\sigma_2}}\|u\|_{H^{s-\sigma_2}}\\
&\quad + \|\nabla V\|_{L^\infty}\big(\|u\|_{C^{\sigma_1}}\|v\|_{H^{s-\sigma_1}} + \|v\|_{L^\infty}\|u\|_{H^s}\big) + \|u\|_{C^{\sigma_1}}\|v\|_{L^\infty}\|V\|_{H^{s+\sigma}}\big),
\end{aligned}$$

where $\|\overline{\partial}_t u\|_{C^{\sigma_1}}\|v\|_{H^{s-\sigma_1}}$ and $\|u\|_{C^{\sigma_1}}\|v\|_{H^{s-\sigma_1}}$ can also be replaced by $\|\overline{\partial}_t u\|_{L^\infty}\|v\|_{H^s}$ and $\|u\|_{L^\infty}\|v\|_{H^s}$ respectively.

PROOF. A direct calculation gives

$$\begin{aligned}
\overline{\partial}_t R(u,v) &= \sum_{|k-\ell|\le 2}\overline{\partial}_t(\Delta_k u\Delta_\ell V)\\
&= \sum_{|k-\ell|\le 2}\Big(\Delta_k(\overline{\partial}_t u)\Delta_\ell v + \Delta_k u\Delta_\ell(\overline{\partial}_t v)\\
&\qquad - [\Delta_k, V]\cdot\nabla u\Delta_\ell v - \Delta_k u[\Delta_\ell, V]\cdot\nabla v\Big)\\
&= R(\overline{\partial}_t u, v) + R(u,\overline{\partial}_t v) - \sum_{|k-\ell|\le 2}\big([\Delta_k, V]\cdot\nabla u\Delta_\ell v + \Delta_k u[\Delta_\ell, V]\cdot\nabla v\big).
\end{aligned}$$

It follows from Lemma 2.10 that

$$\|R(\overline{\partial}_t u, v)\|_{H^s} \le C\|\overline{\partial}_t u\|_{C^{\sigma_1}}\|v\|_{H^{s-\sigma_1}},$$
$$\|R(u, \overline{\partial}_t v)\|_{H^s} \le C\|\overline{\partial}_t v\|_{C^{\sigma_2}}\|u\|_{H^{s-\sigma_2}}.$$

Given $N_0 \in \mathbf{N}$ sufficiently large, we further decompose

$$\begin{aligned}\sum_{|k-\ell|\le 2} [\Delta_k, V]\cdot\nabla u\Delta_\ell v = &\sum_{|k-\ell|\le 2} [\Delta_k, S_{k-N_0}V]\cdot\nabla u\Delta_\ell v\\ &+ \sum_{|k-\ell|\le 2} [\Delta_k, S^{k-N_0}V]\cdot\nabla u\Delta_\ell v,\end{aligned}$$

where $S^k = 1 - S_k$. By noting that u has to be spectrally supported in $\{|\xi|\sim 2^k\}$ and using the commutator estimate

$$\|[\Delta_k, u]\nabla v\|_{L^p} \le C\|\nabla u\|_{L^\infty}\|v\|_{L^p} \quad \text{for any } p\in[1,\infty], \tag{2.6}$$

which follows from the identity

$$\begin{aligned}[\Delta_k, u]\nabla v(x) &= \int_{\mathbf{R}^d} \check{\varphi}_k(x-x')(u(x')-u(x))\nabla v(x')dx'\\ &= \int_{\mathbf{R}^d} \nabla\check{\varphi}_k(x-x')(u(x')-u(x))v(x')dx'\\ &\quad - \int_{\mathbf{R}^d} \check{\varphi}_k(x-x')\nabla u(x')v(x')dx',\end{aligned}$$

and $\|\check{\varphi}_k\|_{L^1} + \|x\nabla\check{\varphi}_k\|_{L^1} \le C$, then we can deduce that

$$\begin{aligned}\big\|[\Delta_k, S_{k-N_0}V]\cdot\nabla u\Delta_\ell v\big\|_{L^2} \le &C2^{-k\sigma_1}\|\nabla V\|_{L^\infty}\|u\|_{C^{\sigma_1}}\|\Delta_\ell v\|_{L^2}\\ \le &Cc_\ell 2^{-\ell s}\|\nabla V\|_{L^\infty}\|u\|_{C^{\sigma_1}}\|v\|_{H^{s-\sigma_1}}\end{aligned}$$

with $\|\{c_\ell\}\|_{\ell^2}\le 1$. Obviously, the term $[\Delta_k, S_{k-N_0}V]\cdot\nabla u\Delta_\ell v$ is spectrally supported in a ball $\{|\xi|\lesssim 2^k\}$ for $|k-\ell|\le 2$. Then Lemma 2.4 ensures that

$$\Big\|\sum_{|k-\ell|\le 2}[\Delta_k, S_{k-N_0}V]\cdot\nabla u\Delta_\ell v\Big\|_{H^s} \le C\|\nabla V\|_{L^\infty}\|u\|_{C^{\sigma_1}}\|v\|_{H^{s-\sigma_1}}.$$

Noticing that

$$\begin{aligned}&\|\Delta_k(S^{k-N_0}V\cdot\nabla u)\Delta_\ell v\|_{L^2} \le \|\Delta_k(S^{k-N_0}V\cdot\nabla S_k u)\Delta_\ell v\|_{L^2}\\ &\qquad + \|\Delta_k(S^{k-N_0}\nabla\cdot VS^k u)\Delta_\ell v\|_{L^2} + \|\Delta_k\nabla\cdot(S^{k-N_0}VS^k u)\Delta_\ell v\|_{L^2}\\ &\le C2^{-\sigma_1 k}\|\nabla V\|_{L^\infty}\|u\|_{C^{\sigma_1}}\|\Delta_\ell v\|_{L^2}.\end{aligned}$$

we can deduce from Lemma 2.4 that

$$\Big\|\sum_{|k-\ell|\le 2}\Delta_k(S^{k-N_0}V\cdot\nabla u)\Delta_\ell v\Big\|_{H^s} \le C\|\nabla V\|_{L^\infty}\|u\|_{C^{\sigma_1}}\|v\|_{H^{s-\sigma_1}}.$$

On the other hand, we have

$$\begin{aligned}\Big\|\sum_{|k-\ell|\le 2} S^{k-N_0}V\cdot\nabla\Delta_k u\Delta_\ell v\Big\|_{H^s} &\le C\sum_{|k-\ell|\le 2}\|S^{k-N_0}V\cdot\nabla\Delta_k u\Delta_\ell v\|_{H^s}\\ &\le C\sum_{|k-\ell|\le 2}2^{(1-\sigma_1)k}\|S^{k-N_0}V\|_{H^s}\|u\|_{C^{\sigma_1}}\|v\|_{L^\infty}\\ &\le C\sum_{|k-\ell|\le 2}2^{-k(\sigma-1+\sigma_1)}\|V\|_{H^{s+\sigma}}\|u\|_{C^{\sigma_1}}\|v\|_{L^\infty}\\ &\le C\|V\|_{H^{s+\sigma}}\|u\|_{C^{\sigma_1}}\|v\|_{L^\infty}.\end{aligned}$$

This shows that

$$\begin{aligned}\Big\|\sum_{|k-\ell|\le 2}[\Delta_k, S^{k-N_0}V]\cdot\nabla u\Delta_\ell v\Big\|_{H^s} \le & C\big(\|\nabla V\|_{L^\infty}\|u\|_{C^{\sigma_1}}\|v\|_{H^{s-\sigma_1}}\\ &+\|V\|_{H^{s+\sigma}}\|u\|_{C^{\sigma_1}}\|v\|_{L^\infty}\big).\end{aligned}$$

Obviously, $\|u\|_{C^{\sigma_1}}\|v\|_{H^{s-\sigma_1}}$ can be replaced by $\|u\|_{L^\infty}\|v\|_{H^s}$ in the above proof.

In a similar way, we can deduce that

$$\begin{aligned}\Big\|\sum_{|k-\ell|\le 2}\Delta_k u[\Delta_\ell, V]\cdot\nabla v\Big\|_{H^s} \le & C\big(\|\nabla V\|_{L^\infty}\|v\|_{L^\infty}\|u\|_{H^s}\\ &+\|V\|_{H^{s+\sigma}}\|v\|_{L^\infty}\|u\|_{C^{\sigma_1}}\big).\end{aligned}$$

This completes the proof of the lemma. □

2.5. Tame estimates in Chemin-Lerner spaces

Let us recall the following lemmas from [**33**].

LEMMA 2.16. *Let $s\in\mathbf{R}$ and $q,q_1,q_2\in[1,\infty]$ with $\frac{1}{q}=\frac{1}{q_1}+\frac{1}{q_2}$. Then for any $s_1>0$, we have*

$$\|T_g f\|_{\widetilde{L}^q_z(I;H^s)}\le C\min\big(\|g\|_{\widetilde{L}^{q_1}_z(I;L^\infty)}\|f\|_{\widetilde{L}^{q_2}_z(I;H^s)}, \|g\|_{\widetilde{L}^{q_1}_z(I;C^{-s_1})}\|f\|_{\widetilde{L}^{q_2}_z(I;H^{s+s_1})}\big).$$

LEMMA 2.17. *Let $s\in\mathbf{R}$ and $q,q_1,q_2,r\in[1,\infty]$ with $\frac{1}{q}=\frac{1}{q_1}+\frac{1}{q_2}$. Then for any $s_1>0$, we have*

$$\|T_g f\|_{\widetilde{L}^q_z(I;B^s_{\infty,r})}\le C\min\big(\|g\|_{\widetilde{L}^{q_1}_z(I;L^\infty)}\|f\|_{\widetilde{L}^{q_2}_z(I;B^s_{\infty,r})}, \|g\|_{\widetilde{L}^{q_1}_z(I;C^{-s_1})}\|f\|_{\widetilde{L}^{q_2}_z(I;B^{s+s_1}_{\infty,r})}\big).$$

LEMMA 2.18. *Let $q,q_1,q_2,r\in[1,\infty]$ with $\frac{1}{q}=\frac{1}{q_1}+\frac{1}{q_2}$. Then for any $s>0$ and $s_1\in R$, we have*

$$\begin{aligned}\|R(f,g)\|_{\widetilde{L}^q_z(I;H^s)}&\le C\|g\|_{\widetilde{L}^{q_1}_z(I;C^{s_1})}\|f\|_{\widetilde{L}^{q_2}_z(I;H^{s-s_1})},\\ \|R(f,g)\|_{\widetilde{L}^q_z(I;B^s_{\infty,r})}&\le C\|g\|_{\widetilde{L}^{q_1}_z(I;C^{s_1})}\|f\|_{\widetilde{L}^{q_2}_z(I;B^{s-s_1}_{\infty,r})}.\end{aligned}$$

If $s\le 0$ and $s_1+s_2>0$, then we have

$$\begin{aligned}\|R(f,g)\|_{\widetilde{L}^q_z(I;H^s)}&\le C\|g\|_{\widetilde{L}^{q_1}_z(I;C^{s_1})}\|f\|_{\widetilde{L}^{q_2}_z(I;H^{s_2})},\\ \|R(f,g)\|_{\widetilde{L}^q_z(I;B^s_{\infty,r})}&\le C\|g\|_{\widetilde{L}^{q_1}_z(I;C^{s_1})}\|f\|_{\widetilde{L}^{q_2}_z(I;C^{s_2})}.\end{aligned}$$

2.6. Commutator estimates

LEMMA 2.19. *Let $m, \mu \in \mathbf{R}, s > 0$ and $a \in \Gamma_\rho^m(\mathbf{R}^d)$ with $\rho \in (0,1]$. Then there holds*

$$\big\|[\langle D\rangle^s, T_a]u\big\|_{H^\mu} \le CM_\rho^m(a)\|u\|_{H^{s+\mu+m-\rho}}.$$

PROOF. We write

$$[\langle D\rangle^s, T_a]u = T_{\langle\xi\rangle^s}T_a u - T_{\langle\xi\rangle^s a}u + (\langle D\rangle^s - T_{\langle\xi\rangle^s})T_a u,$$

then the proposition follows from Proposition 2.6 and the fact that

$$\|(\langle D\rangle^s - T_{\langle\xi\rangle^s})T_a u\|_{H^\mu} \le C\|\langle D\rangle^s(1-\psi(D))T_a u\|_{H^\mu} \le CM_0^m(a)\|u\|_{H^{-\mu'}}$$

for any $\mu' > 0$. □

LEMMA 2.20. *Let $\mathcal{S} = \mathbf{R}^d \times I$ with $I \subset \mathbf{R}$ an interval. Then it holds that for any integer $s \ge 1$,*

$$\|[\nabla_{x,y}^s, v]\nabla_{x,y}u\|_{L^2(\mathcal{S})} \le C\big(\|\nabla_{x,y}v\|_{L^\infty(\mathcal{S})}\|u\|_{H^s(\mathcal{S})} + \|u\|_{L^\infty(\mathcal{S})}\|\nabla_{x,y}v\|_{H^s(\mathcal{S})}\big).$$

PROOF. When $\mathcal{S} = \mathbf{R}^{d+1}$, this lemma is classical. General case can be deduced by using an extension argument. □

PROPOSITION 2.21. *Let $V \in C([0,T]; B^1_{\infty,1}(\mathbf{R}^d))$ and $p = p(t,x,\xi)$ be homogenous in ξ of order m. Then for any $s \ge 0$,*

$$\begin{aligned}&\big\|[T_p, \partial_t + T_V\cdot\nabla]u(t)\big\|_{H^s}\\ &\le C\Big(M_0^m(p)\|V(t)\|_{B^1_{\infty,1}} + M_0^m\big((\partial_t + T_V\cdot\nabla)p\big)\Big)\|u(t)\|_{H^{s+m}}.\end{aligned}$$

If $p(t) \in \Gamma_\mu^m$ for some $\mu > 0$, $M_0^m(p)\|V(t)\|_{B^1_{\infty,1}}$ can be replaced by $M_\mu^m(p)\|V(t)\|_{W^{1,\infty}}$ in the case of $s > 0$.

PROOF. The proof is motivated by Lemma 2.16 in [**3**]. As in [**3**], it suffices to consider the case when $p = p(t,x)$ by decomposing p into a sum of spherical harmonic. In this case, $M_0^0(p) = \|p\|_{L^\infty}$. A direct calculation gives

$$\begin{aligned}[\partial_t + T_V\cdot\nabla, T_p]u =& T_{\partial_t p}u + T_V\cdot T_{\nabla p}u + T_V\cdot T_p\nabla u - T_pT_V\cdot\nabla u\\ =& T_{(\partial_t + T_V\cdot\nabla)p}u - \big(T_{T_V\cdot\nabla p} - T_V\cdot T_{\nabla p}\big)u\\ &+ \big(T_V\cdot T_p\nabla u - T_pT_V\cdot\nabla u\big).\end{aligned}$$

We infer from Lemma 2.10 that

$$\|T_{(\partial_t+T_V\cdot\nabla)p}u\|_{H^s} \le C\|(\partial_t + T_V\cdot\nabla)p\|_{L^\infty}\|u\|_{H^s}. \tag{2.7}$$

Recalling the definition $S^j u = u - S_j u$, we decompose $T_V \cdot T_{\nabla p} u$ as

$$\begin{aligned}
T_V \cdot T_{\nabla p} u &= \sum_{j,k} S_{j-3}V \cdot \Delta_j\big(S_{k-3}(\nabla p)\Delta_k u\big) \\
&= \sum_k S_{k-3}V \cdot S_{k-3}(\nabla p)\Delta_k u + \sum_{j,k} \big(S_{j-3}V - S_{k-3}V\big) \cdot \Delta_j\big(S_{k-3}(\nabla p)\Delta_k u\big) \\
&= \sum_k S_{k-3}(V \cdot \nabla p)\Delta_k u - \sum_k S_{k-3}(S^{k-3}V \cdot \nabla p)\Delta_k u \\
&\quad - \sum_k [S_{k-3}, S_{k-3}V] \cdot \nabla p \Delta_k u + \sum_{j,k} \big(S_{j-3}V - S_{k-3}V\big) \cdot \Delta_j\big(S_{k-3}(\nabla p)\Delta_k u\big) \\
&\triangleq T_{V\cdot\nabla p} u + I_1 + I_2 + I_3.
\end{aligned}$$

We get by Lemma 2.2 that

$$\begin{aligned}
&\|\big(S_{j-3}V - S_{k-3}V\big) \cdot \Delta_j\big(S_{k-3}(\nabla p)\Delta_k u\big)\|_{L^2} \\
&\quad \le C2^{-k(s-1)}\|S_{j-3}V - S_{k-3}V\|_{L^\infty}\|p\|_{L^\infty}\|u\|_{H^s}.
\end{aligned}$$

Note that the summation index (j,k) in I_3 satisfies $|k-j| \le N_0$ for some $N_0 \in \mathbf{N}$, and $\big(S_{j-3}V - S_{k-3}V\big) \cdot \Delta_j\big(S_{k-3}(\nabla p)\Delta_k u\big)$ is spectrally supported in a ball $\{|\xi| \lesssim 2^j\}$. Then Lemma 2.4 ensures that for $s > 0$,

$$\|I_3\|_{H^s} \le C\|V\|_{W^{1,\infty}}\|p\|_{L^\infty}\|u\|_{H^s}$$

by using the fact that

$$\sum_{|j-k|\le N_0} 2^k\|S_{j-3}V - S_{k-3}V\|_{L^\infty} \le C\|V\|_{W^{1,\infty}}.$$

For $s = 0$, we have

$$\|I_3\|_{L^2} \le C\|V\|_{B^1_{\infty,1}}\|p\|_{L^\infty}\|u\|_{L^2}.$$

By Lemma 2.2 again, we have

$$\begin{aligned}
\|S_{k-3}(S^{k-3}V \cdot \nabla p)\Delta_k u\|_{L^2} &\le C2^{-ks}c_k\big(\|S^{k-3}(\nabla V)\|_{L^\infty} + 2^k\|S^{k-3}V\|_{L^\infty}\big)\|p\|_{L^\infty}\|u\|_{H^s} \\
&\le C2^{-ks}c_k\|V\|_{W^{1,\infty}}\|p\|_{L^\infty}\|u\|_{H^s},
\end{aligned}$$

with $\|\{c_k\}\|_{\ell^2} \le 1$, and $S_{k-3}(S^{k-3}V \cdot \nabla p)\Delta_k u$ is spectrally supported in an annulus $\{|\xi| \sim 2^k\}$. Then Lemma 2.4 ensures that

$$\|I_1\|_{H^s} \le C\|p\|_{L^\infty}\|V\|_{W^{1,\infty}}\|u\|_{H^s}.$$

Noticing that $[S_{k-3}, S_{k-3}V] \cdot \nabla p \Delta_k u$ is spectrally supported in a ball $\{|\xi| \lesssim 2^k\}$ and by (2.6),

$$\|[S_{k-3}, S_{k-3}V] \cdot \nabla p \Delta_k u\|_{L^2} \le Cc_k 2^{-ks}\|V\|_{W^{1,\infty}}\|p\|_{L^\infty}\|u\|_{H^s},$$

with $\|\{c_k\}\|_{\ell^2} \le 1$, we infer from Lemma 2.4 that for $s > 0$,

$$\|I_2\|_{H^s} \le C\|p\|_{L^\infty}\|V\|_{W^{1,\infty}}\|u\|_{H^s}.$$

In the case of $s = 0$, we need to decompose I_2 as

$$I_2 = \sum_k [S_{k-3}, S_{k-N_0}V] \cdot \nabla p \Delta_k u + \sum_k [S_{k-3}, (S_{k-3} - S_{k-N_0})V] \cdot \nabla p \Delta_k u,$$

where we take N_0 big enough so that $[S_{k-3}, S_{k-N_0}V]\cdot\nabla p\Delta_k u$ is spectrally supported in an annulus $\{|\xi|\sim 2^j\}$. Then we have

$$\|I_2\|_{L^2}\le C\|p\|_{L^\infty}\|V\|_{B^1_{\infty,1}}\|u\|_{L^2}.$$

Putting the estimates of I_1, I_2 and I_3 together, we deduce that

$$\|(T_V\cdot T_{\nabla p}-T_{V\cdot\nabla p})u\|_{H^s}\le C\|p\|_{L^\infty}\|V\|_{W^{1,\infty}}\|u\|_{H^s}(s>0),$$
$$\|(T_V\cdot T_{\nabla p}-T_{V\cdot\nabla p})u\|_{L^2}\le C\|p\|_{L^\infty}\|V\|_{B^1_{\infty,1}}\|u\|_{L^2}.$$

By Bony's decomposition (2.3), we have

$$\begin{aligned}\|T_{V\cdot\nabla p-T_V\cdot\nabla p}u\|_{H^s}&\le\|V\cdot\nabla p-T_V\cdot\nabla p\|_{L^\infty}\|u\|_{H^s}\\&\le C\|p\|_{L^\infty}\|V\|_{B^1_{\infty,1}}\|u\|_{H^s},\end{aligned}$$

and if $p\in C^\mu$, we have

$$\|T_{V\cdot\nabla p-T_V\cdot\nabla p}u\|_{H^s}\le C\|p\|_{C^\mu}\|V\|_{W^{1,\infty}}\|u\|_{H^s}.$$

This shows that for $s>0$

$$\|(T_{T_V\cdot\nabla p}-T_V\cdot T_{\nabla p})u\|_{H^s}\le C\min\big(\|p\|_{L^\infty}\|V\|_{B^1_{\infty,1}},\|p\|_{C^\mu}\|V\|_{W^{1,\infty}}\big)\|u\|_{H^s},\tag{2.8}$$

and for $s=0$,

$$\|(T_{T_V\cdot\nabla p}-T_V\cdot T_{\nabla p})u\|_{H^s}\le\|p\|_{L^\infty}\|V\|_{B^1_{\infty,1}}\|u\|_{H^s}.\tag{2.9}$$

Next we decompose $T_V\cdot T_p\nabla u$ as

$$\begin{aligned}T_V\cdot T_p\nabla u&=\sum_{j,k}S_{j-3}V\Delta_j\big(S_{k-3}p\cdot\Delta_k\nabla u\big)\\&=\sum_k S_{k-3}VS_{k-3}p\cdot\Delta_k\nabla u\\&\quad+\sum_{j,k}\big(S_{j-3}V-S_{k-3}V\big)\Delta_j\big(S_{k-3}p\cdot\Delta_k\nabla u\big)\\&\triangleq\sum_k S_{k-3}VS_{k-3}p\cdot\Delta_k\nabla u+II_1.\end{aligned}$$

On the other hand, we have

$$\begin{aligned}T_pT_V\cdot\nabla u&=\sum_j S_{j-3}p\Delta_j\big(S_{j-3}V\cdot\nabla u\big)\\&\quad+\sum_{j,k}S_{j-3}p\Delta_j\big((S_{k-3}-S_{j-3})V\cdot\nabla\Delta_k u\big)\\&=\sum_j S_{j-3}VS_{j-3}p\cdot\Delta_j\nabla u+\sum_j S_{j-3}p[\Delta_j,S_{j-3}V]\cdot\nabla u\\&\quad+\sum_{j,k}S_{j-3}p\cdot\Delta_j\big((S_{k-3}-S_{j-3})V\cdot\nabla\Delta_k u\big)\\&\triangleq\sum_j S_{j-3}VS_{j-3}p\cdot\nabla\Delta_j u+II_2+II_3.\end{aligned}$$

This gives

$$T_V\cdot T_p\nabla u-T_pT_V\cdot\nabla u=II_1-II_2-II_3.$$

Notice that the summation index (j,k) in II_3 satisfies $|k-j|\le N_0$ for some $N_0\in\mathbf{N}$. Similar to I_3, we can deduce that

$$\|II_3\|_{H^s}\le C\|p\|_{L^\infty}\|V\|_{W^{1,\infty}}\|u\|_{H^s}.$$

For II_1, we have

$$\|II_1\|_{H^s} \le C\|p\|_{L^\infty}\|V\|_{W^{1,\infty}}\|u\|_{H^s}(s>0),$$
$$\|II_1\|_{L^2} \le C\|p\|_{L^\infty}\|V\|_{B^1_{\infty,1}}\|u\|_{L^2}.$$

We decompose II_2 as

$$\begin{aligned}II_2 =&\sum_j S_{j-3}p\big(\Delta_j(T_{S_{j-3}V}\cdot\nabla u) - T_{S_{j-3}V}\cdot\nabla\Delta_j u\big)\\ &+\sum_j S_{j-3}p\Delta_j\big(S_{j-3}V\cdot\nabla u - T_{S_{j-3}V}\cdot\nabla u\big)\\ &-\sum_j S_{j-3}p\big(S_{j-3}V\cdot\nabla\Delta_j u - T_{S_{j-3}V}\cdot\nabla\Delta_j u\big)\\ \triangleq& II_2^1 + II_2^2 + II_2^3.\end{aligned}$$

Similar to I_2, we have

$$\|II_2^1\|_{H^s} \le C\|p\|_{L^\infty}\|V\|_{W^{1,\infty}}\|u\|_{H^s}(s>0),$$
$$\|II_2^1\|_{L^2} \le C\|p\|_{L^\infty}\|V\|_{B^1_{\infty,1}}\|u\|_{L^2}.$$

Using Bony's decomposition (2.3), we decompose II_2^2 as

$$II_2^2 = \sum_j S_{j-3}p\Delta_j\big(R(S_{j-3}V,\nabla u) + T_{S^{j-N_0}\nabla u}\cdot S_{j-3}V\big)$$

for some $N_0\in\mathbf{N}$. It is easy to prove that for $s>0$,

$$\|S_{j-3}p\Delta_j\big(R(S_{j-3}V,\nabla u) + T_{S^{j-N_0}\nabla u}\cdot S_{j-3}V\big)\|_{L^2} \le Cc_j2^{-js}\|p\|_{L^\infty}\|V\|_{W^{1,\infty}}\|u\|_{H^s}$$

with $\|\{c_j\}\|_{\ell^2}\le 1$. Then Lemma 2.4 ensures that for $s>0$,

$$\|II_2^2\|_{H^s} \le C\|p\|_{L^\infty}\|V\|_{W^{1,\infty}}\|u\|_{H^s}.$$

For $s=0$, it is obvious that

$$\|II_2^2\|_{L^2} \le C\|p\|_{L^\infty}\|V\|_{B^1_{\infty,1}}\|u\|_{L^2}.$$

Similarly, we have

$$\|II_2^3\|_{H^s} \le C\|p\|_{L^\infty}\|V\|_{W^{1,\infty}}\|u\|_{H^s}(s>0),$$
$$\|II_2^3\|_{L^2} \le C\|p\|_{L^\infty}\|V\|_{B^1_{\infty,1}}\|u\|_{L^2}.$$

This proves that

$$\|T_V\cdot T_p\nabla u - T_pT_V\cdot\nabla u\|_{H^s} \le C\|p\|_{L^\infty}\|V\|_{W^{1,\infty}}\|u\|_{H^s}(s>0),$$
$$\|T_V\cdot T_p\nabla u - T_pT_V\cdot\nabla u\|_{L^2} \le C\|p\|_{L^\infty}\|V\|_{B^1_{\infty,1}}\|u\|_{L^2},$$

which together with (2.7)-(2.9) give the proposition. □

REMARK 2.22. If the symbol $p(t,x,\xi)$ satisfies

$$M^m_{-\mu}(\partial_t + T_V\cdot\nabla p) \triangleq \sup_{|\alpha|\le 3d/2+1}\sup_{|\xi|\ge 1/2}\|(1+|\xi|)^{|\alpha|-m}\partial_\xi^\alpha(\partial_t + T_V\cdot\nabla p)\|_{C^{-\mu}} < \infty$$

for some $\mu>0$, then there holds

$$\begin{aligned}&\big\|[T_p,\partial_t + T_V\cdot\nabla]u(t)\big\|_{H^s}\\ &\le CM^m_0(p)\|V(t)\|_{B^1_{\infty,1}}\|u(t)\|_{H^{s+m}} + CM^m_{-\mu}\big((\partial_t + T_V\cdot\nabla)p\big)\|u(t)\|_{H^{s+m+\mu}},\end{aligned}$$

which can be seen from (2.7).

The following proposition will be used in the proof of well-posedness. Although the estimate is very rough, it is sufficient for our application.

PROPOSITION 2.23. *Assume that $V \in C([0,T];W^{2,\infty}), \partial_t V \in C([0,T];W^{1,\infty})$, and the symbol $p = p(t,x,\xi)$ is homogenous in ξ of order $m \in \mathbf{R}$. Then it holds that*

$$\begin{aligned}\big\|(\partial_t + T_V\cdot\nabla)[T_p, \partial_t + T_V\cdot\nabla]u(t)\big\|_{H^s} &\le C\big(1 + \|V(t)\|_{W^{2,\infty}} + \|\partial_t V(t)\|_{W^{1,\infty}}\big)^2 \\ &\quad\times \big(M_2^m(p) + M_1^m(\partial_t p) + M_0^m(\partial_t^2 p)\big)\|\big((\partial_t + T_V\cdot\nabla)u(t), u(t)\big)\|_{H^{m+s}}.\end{aligned}$$

PROOF. Note that

$$[T_p, \partial_t + T_V\cdot\nabla]u(t) = -\big(T_{\partial_t p} + T_V T_{\nabla p}\big)u - \big(T_V T_p - T_p T_V\big)\nabla u.$$

A direct calculation gives

$$\begin{aligned}&(\partial_t + T_V\cdot\nabla)[T_p, \partial_t + T_V\cdot\nabla]u(t) \\ &= -(\partial_t + T_V\cdot\nabla)\big((T_{\partial_t p} + T_V T_{\nabla p})u + (T_V T_p - T_p T_V)\nabla u\big) \\ &= -\big[\partial_t + T_V\cdot\nabla, T_{\partial_t p} + T_V T_{\nabla p}\big]u + (T_{\partial_t p} + T_V T_{\nabla p})(\partial_t + T_V\cdot\nabla)u \\ &\quad -\big[\partial_t + T_V\cdot\nabla, (T_V T_p - T_p T_V)\big](\nabla u) - [T_p, T_V](\partial_t + T_V\cdot\nabla)(\nabla u).\end{aligned}$$

It follows from Lemma 2.10 that

$$\begin{aligned}&\big\|(T_{\partial_t p} + T_V T_{\nabla p})(\partial_t + T_V\cdot\nabla)u\big\|_{H^s} \\ &\le C\big(M_0^m(\partial_t p) + \|V(t)\|_{L^\infty} M_1^m(p)\big)\|(\partial_t + T_V\cdot\nabla)u\|_{H^{m+s}}.\end{aligned}$$

We decompose

$$\begin{aligned}&\big[\partial_t + T_V\cdot\nabla, T_{\partial_t p} + T_V T_{\nabla p}\big]u \\ &= \big[\partial_t + T_V\cdot\nabla, T_{\partial_t p}\big]u + \big[\partial_t + T_V\cdot\nabla, T_V\big]T_{\nabla p}u + T_V\big[\partial_t + T_V\cdot\nabla, T_{\nabla p}\big]u.\end{aligned}$$

Then it follows from Proposition 2.21 and Proposition 2.6 that

$$\begin{aligned}&\big\|\big[\partial_t + T_V\cdot\nabla, T_{\partial_t p} + T_V T_{\nabla p}\big]u\|_{H^s} \\ &\le C\big(1 + \|V(t)\|_{C^{1+\varepsilon}} + \|\partial_t V\|_{L^\infty}\big)^2\big(M_0^m(\nabla_{t,x}p) + M_0^m(\nabla_{t,x}^2 p)\big)\|u(t)\|_{H^{m+s}}.\end{aligned}$$

We decompose

$$[T_p, T_V](\partial_t + T_V\cdot\nabla)(\nabla u) = [T_p, T_V]\nabla(\partial_t + T_V\cdot\nabla)u - [T_p, T_V]T_{\nabla V}\cdot\nabla u,$$

which along with Proposition 2.6 gives

$$\begin{aligned}&\big\|[T_p, T_V](\partial_t + T_V\cdot\nabla)(\nabla u)\big\|_{H^s} \\ &\le C\|V(t)\|_{W^{1,\infty}} M_1^m(p)\big(\|(\partial_t + T_V\cdot\nabla)u\|_{H^{m+s}} + \|\nabla V(t)\|_{L^\infty}\|u\|_{H^{m+s}}\big).\end{aligned}$$

We write

$$\big[\partial_t, (T_V T_p - T_p T_V)\big](\nabla u) = [T_{\partial_t V}, T_p](\nabla u) + [T_V, T_{\partial_t p}](\nabla u),$$

from which and Proposition 2.6, we deduce that

$$\big\|[\partial_t, (T_V T_p - T_p T_V)](\nabla u)\big\|_{H^s} \le C\|(V, \partial_t V)\|_{W^{1,\infty}}\big(M_1^m(p) + M_1^m(\partial_t p)\big)\|u\|_{H^{m+s}}.$$

Finally, we deduce from Corollary 2.9 that

$$\big\|\big[T_V\cdot\nabla, [T_V, T_p]\big](\nabla u)\big\|_{H^s} \le C\|V\|_{W^{2,\infty}} M_2^m(p)\|u\|_{H^{m+s}}.$$

Putting the above estimates together, the proposition is proved. □

CHAPTER 3

Parabolic evolution equation

Let $I=[z_0,z_1]$ be a finite interval. We denote by $\Gamma^m_\rho(I\times\mathbf{R}^d)$ the space of symbols $a(z;x,\xi)$ satisfying

$$\mathcal{M}^m_\rho(a)\overset{\text{def}}{=}\sup_{z\in I}\sup_{|\alpha|\le 3d/2+1+\rho}\sup_{|\xi|\ge 1/2}\|(1+|\xi|)^{|\alpha|-m}\partial^\alpha_\xi a(z;\cdot,\xi)\|_{W^{\rho,\infty}}<+\infty.$$

We consider the parabolic evolution equation

$$\begin{cases}\partial_z w+T_a w=f,\\ w|_{z=z_0}=w_0,\end{cases}\tag{3.1}$$

where the symbol $a\in\Gamma^1_\rho(I\times\mathbf{R}^d)$ is elliptic in the sense that there exists $c_1>0$ such that for any $z\in I,(x,\xi)\in\mathbf{R}^d\times\mathbf{R}^d$, it holds that

$$\operatorname{Re}a(z;x,\xi)\ge c_1|\xi|.\tag{3.2}$$

For the elliptic estimates, we need to introduce two kinds of spaces. The first kind of spaces are intended for Sobolev elliptic estimates:

$$X^\sigma(I)\overset{\text{def}}{=}\widetilde{L}^\infty_z(I;H^\sigma(\mathbf{R}^d))\cap L^2_z(I;H^{\sigma+\frac12}(\mathbf{R}^d)),$$
$$Y^\sigma(I)\overset{\text{def}}{=}\widetilde{L}^1_z(I;H^\sigma(\mathbf{R}^d))+L^2_z(I;H^{\sigma-\frac12}(\mathbf{R}^d)).$$

The second kind of spaces are intended for Hölder elliptic estimates:

$$X^\sigma_{p,q}(I)\overset{\text{def}}{=}\widetilde{L}^p_z(I;B^{\sigma+\frac1p}_{\infty,q}),$$
$$Y^\sigma_q(I)\overset{\text{def}}{=}\widetilde{L}^1_z(I;B^\sigma_{\infty,q})+\widetilde{L}^2_z(I;B^{\sigma-\frac12}_{\infty,q})+\widetilde{L}^\infty_z(I;B^{\sigma-1}_{\infty,q}).$$

Let us recall the following parabolic estimates in Chemin-Lerner type spaces, which have been essentially proved in [**33**].

PROPOSITION 3.1. *Let $a\in\Gamma^1_\rho(I\times\mathbf{R}^d)$ for some $\rho>0$. Assume that $w_0\in H^\sigma$ and $f\in Y^\sigma(I)$ for $\sigma\in\mathbf{R}$. Then there exists a unique solution $w\in X^\sigma(I)$ of (3.1) satisfying*

$$\|w\|_{X^\sigma(I)}\le C\big(\mathcal{M}^1_\rho(a)\big)\big(\|w_0\|_{H^\sigma}+\|f\|_{Y^\sigma(I)}\big),$$

where C is an increasing function depending on c_1 and $|I|$.

PROPOSITION 3.2. *Let $\sigma\in\mathbf{R},p,q\in[1,\infty]$ and $a\in\Gamma^1_\rho(I\times\mathbf{R}^d)$ for some $\rho>0$. Assume that $w\in X^\sigma_{p,q}(I)$ is a solution of (3.1). Then for any $\delta>0$,*

$$\|w\|_{X^\sigma_{p,q}(I)}\le C\big(\mathcal{M}^1_\rho(a)\big)\big(\|w_0\|_{B^\sigma_{\infty,q}}+\|f\|_{Y^\sigma_{p,q}(I)}+\|w\|_{\widetilde{L}^p_z(I;C^{-\delta})}\big),$$

where $Y^\sigma_{p,q}(I)=\widetilde{L}^p_z(I;B^{\sigma-1+\frac1p}_{\infty,q})$ and C is an increasing function depending on c_1 and $|I|$.

CHAPTER 4

Elliptic estimates in a strip

The goal of this chapter is to establish the elliptic estimates in Sobolev spaces and Besov spaces. These estimates will be used to estimate the Dirichlet-Neumann operator, the velocity and the pressure.

Throughout this chapter, we assume that $\eta \in C^{\frac{3}{2}+\varepsilon}(\mathbf{R}^d)$ for some $\varepsilon > 0$ and there exists some $h_0 > 0$ such that

$$1 + \eta(x) \geq h_0 \quad \text{for } x \in \mathbf{R}^d. \tag{4.1}$$

Let $\mathcal{S} \triangleq \{(x,y) : x \in \mathbf{R}^d, -1 < y < \eta(x)\}$ be a strip.

In the sequel, we denote by $K_\eta = C(\|\eta\|_{C^{\frac{3}{2}+\varepsilon}})$ an increasing function depending on h_0, which may change from line to line; $\nabla = (\partial_{x_1}, \cdots, \partial_{x_d})$, $D = \frac{\nabla}{i}$ and $\Delta = \Delta_x$.

4.1. Elliptic boundary problem

We consider the elliptic equations in $\mathcal{S}$:

$$\Delta_{x,y}\phi = g \quad \text{in} \quad \mathcal{S}, \tag{4.2}$$

with the Dirichlet boundary conditions

$$\phi|_{y=\eta} = f, \quad \phi|_{y=-1} = f_b, \tag{4.3}$$

or the mixed boundary conditions

$$\phi|_{y=\eta} = f, \quad \partial_y\phi|_{y=-1} = 0. \tag{4.4}$$

Given $f, f_b \in H^{\frac{1}{2}}(\mathbf{R}^d)$ and $g \in H^{-1}(\mathcal{S})$, the existence and uniqueness of weak solutions for the elliptic equations (4.2)-(4.3) or (4.2)-(4.4) can be deduced by using Riesz theorem(see [**3**] for example). Moreover, the solution $\phi \in H^1(\mathcal{S})$ satisfies

$$\|\phi\|_{H^1(\mathcal{S})} \leq C(\|\eta\|_{W^{1,\infty}}, h_0)\big(\|(f, f_b)\|_{H^{\frac{1}{2}}} + \|g\|_{H^{-1}(\mathcal{S})}\big). \tag{4.5}$$

As a warm-up, we first establish the elliptic estimates in the flat strip $\mathcal{S} = [-1, 0] \times \mathbf{R}^d$ by using the parabolic estimates. The same ideas will be used to deal with the general case in the next subsections. In this subsection, we denote that $I = [-1, 0]$.

PROPOSITION 4.1. *Let ϕ be a solution of (4.2)-(4.3). Then for any $\sigma \geq -\frac{1}{2}$, there holds*

$$\|\nabla_{x,y}\phi\|_{X^\sigma(I)} \leq C\big(\|\nabla_{x,y}\phi\|_{L^2(\mathcal{S})} + \|(f, f_b)\|_{H^{\sigma+1}} + \|g\|_{Y^\sigma(I)}\big),$$
$$\|\nabla_{x,y}\phi\|_{H^{\sigma+\frac{1}{2}}(\mathcal{S})} \leq C\big(\|\nabla_{x,y}\phi\|_{L^2(\mathcal{S})} + \|(f, f_b)\|_{H^{\sigma+1}} + \|g\|_{H^{\sigma-\frac{1}{2}}(\mathcal{S})}\big).$$

PROOF. We split the proof into three steps.

Step 1: Estimates on strip domain $[-1/2, 0] \times \mathbf{R}^d$.

Let $w = \chi(y)(\partial_y - |D|)\phi$, where $\chi(y)$ is a smooth function satisfying $\chi(-1) = 0$ and $\chi(y) = 1$ for $y \in [-\frac{1}{2}, 0]$. Then w satisfies

$$(\partial_y + |D|)w = \chi(y)g + \chi'(y)(\partial_y + |D|)\phi \triangleq g_1, \quad w(-1) = 0.$$

Then for any $y \in [-1, 0]$, we have

$$w(x, y) = \int_{-1}^{y} e^{(y'-y)|D|} g_1(x, y')dy'.$$

Using the fact(see [**41**] for example) that there exists $c > 0$ such that for any $y \leq 0$ and $p \in [1, \infty]$, there holds

$$\|e^{y|D|}\Delta_j f\|_{L^p} \leq Ce^{cy2^j}\|\Delta_j f\|_{L^p}, \tag{4.6}$$

we can easily deduce that for $\sigma_1 = \min(\sigma, \frac{1}{2})$,

$$\|w\|_{X^{\sigma_1}(I)} \leq C\big(\|g\|_{Y^\sigma(I)} + \|\nabla_{x,y}\phi\|_{L^2(\mathcal{S})}\big).$$

On the other hand, we have

$$(\partial_y - |D|)\phi = w(y) \quad \text{for } y \in [-\frac{1}{2}, 0], \quad \phi(0) = f.$$

Hence, for $y \in [-\frac{1}{2}, 0]$,

$$\phi(x, y) = e^{y|D|}f + \int_0^y e^{-(y'-y)|D|}w(y')dy',$$

from which and (4.6), we infer that

$$\begin{aligned}\|\nabla_{x,y}\phi\|_{X^{\sigma_1}([-\frac{1}{2},0])} \leq& C\big(\|f\|_{H^{\sigma_1+1}} + \|w\|_{X^{\sigma_1}(I)}\big)\\ \leq& C\big(\|f\|_{H^{\sigma_1+1}} + \|g\|_{Y^\sigma(I)} + \|\nabla_{x,y}\phi\|_{L^2(\mathcal{S})}\big).\end{aligned} \tag{4.7}$$

Step 2: Estimates on strip domain $[-1, -1/2] \times \mathbf{R}^d$.

Let $w = \chi(y)(\partial_y + |D|)\phi$, where $\chi(y)$ is a smooth function satisfying $\chi(0) = 0$ and $\chi(y) = 1$ for $y \in [-1, -\frac{1}{2}]$. Then we have

$$\begin{aligned}&(\partial_y - |D|)w = \chi(y)g + \chi'(y)(\partial_y - |D|)\phi, \quad w(0) = 0,\\ &(\partial_y + |D|)\phi = w \quad \text{for } y \in [-1, -\frac{1}{2}], \quad \phi(-1) = f_b.\end{aligned}$$

Then a similar argument leading to (4.7) gives

$$\|\nabla_{x,y}\phi\|_{X^{\sigma_1}([-1,-\frac{1}{2}])} \leq C\big(\|f_b\|_{H^{\sigma_1+1}} + \|g\|_{Y^\sigma(I)} + \|\nabla_{x,y}\phi\|_{L^2(\mathcal{S})}\big),$$

which along with (4.7) gives

$$\|\nabla_{x,y}\phi\|_{X^{\sigma_1}(I)} \leq C\big(\|(f, f_b)\|_{H^{\sigma_1+1}} + \|g\|_{Y^\sigma(I)} + \|\nabla_{x,y}\phi\|_{L^2(\mathcal{S})}\big). \tag{4.8}$$

A bootstrap argument will ensure that the inequality (4.8) holds for $\sigma_1 = \sigma$.

Step 3: Proof of the second result.

To show the second inequality, it suffices to consider the estimate of the normal derivative. Let $\sigma + \frac{1}{2} = k + \sigma_1$ for $k \in \mathbf{N}$ and $\sigma_1 \in [0, 1)$. Using the interpolation inequality and the elliptic equation, we deduce that

$$\begin{aligned}\|\nabla^k\partial_y\phi\|_{H^{\sigma_1}(\mathcal{S})} \leq& C\big(\|\partial_y\phi\|_{L^2(I;H^{\sigma+\frac{1}{2}})} + \|\partial_y^2\phi\|_{L^2(I;H^{\sigma-\frac{1}{2}})}\big)\\ \leq& C\big(\|\nabla_{x,y}\phi\|_{L^2(I;H^{\sigma+\frac{1}{2}})} + \|g\|_{H^{\sigma-\frac{1}{2}}(\mathcal{S})}\big)\\ \leq& C\big(\|(f, f_b)\|_{H^{\sigma_1+1}} + \|g\|_{H^{\sigma-\frac{1}{2}}(\mathcal{S})} + \|\nabla_{x,y}\phi\|_{L^2(\mathcal{S})}\big).\end{aligned}$$

Let us assume that for $\ell \in [1,k]$, there holds

$$\|\nabla^{k-\ell+1}\partial_y^\ell \phi\|_{H^{\sigma_1}(\mathcal{S})} \le C\big(\|(f,f_b)\|_{H^{\sigma_1+1}} + \|g\|_{H^{\sigma-\frac12}(\mathcal{S})} + \|\nabla_{x,y}\phi\|_{L^2(\mathcal{S})}\big).$$

Then the estimate $\|\partial_y^{k+1}\phi\|_{H^{\sigma_1}(\mathcal{S})}$ can be deduced from

$$\begin{aligned}\|\partial_y^{k+1}\phi\|_{H^{\sigma_1}(\mathcal{S})} \le& C\big(\|\partial_y^{k-1}\Delta\phi\|_{L^2(I;H^{\sigma_1})} + \|g\|_{H^{\sigma-\frac12}(\mathcal{S})}\big)\\ \le& C\big(\|\partial_y\phi\|_{L^2(I;H^{\sigma+\frac12})} + \|\partial_y^k\phi\|_{L^2(I;H^{\sigma_1+1})} + \|g\|_{H^{\sigma-\frac12}(\mathcal{S})}\big),\end{aligned}$$

and an induction assumption. The proof is finished. □

Next, we give the estimates of ϕ in the Besov space.

PROPOSITION 4.2. *Let ϕ be a solution of (4.2)-(4.3) with $g = g_1 + \partial_y g_2$. Let $\sigma \in \mathbf{R}, q, r \in [1,\infty]$. Then it holds that for any $\delta > 0$,*

$$\|\nabla_{x,y}\phi\|_{X^\sigma_{r,q}(I)} \le C\big(\|(f,f_b)\|_{B^{\sigma+1}_{\infty,q}} + \|g_1\|_{Y^\sigma_q(I)} + \|g_2\|_{X^\sigma_{r,q}(I)} + \|\nabla_{x,y}\phi\|_{L^2_z(I;C^{-\delta})}\big).$$

PROOF. Let $w = \chi(y)(\partial_y - |D|)\phi$, where $\chi(y)$ is a smooth function satisfying $\chi(-1) = 0$ and $\chi(y) = 1$ for $y \in [-\frac12, 0]$. Then w satisfies

$$(\partial_y + |D|)(w - \chi(y)g_2) = \chi(y)g_1 + \chi'(y)(\partial_y + |D|)\phi - \chi(y)|D|g_2 - \partial_y\chi g_2 \triangleq g$$

with $w(-1) = 0$. Then for any $y \in [-1,0]$, we have

$$(w - \chi(y)g_2)(x,y) = \int_{-1}^y e^{(y'-y)|D|}g(y')dy'.$$

from which and (4.6), we infer that for $\sigma_1 = \min(-\delta + \frac13, \sigma)$,

$$\|w\|_{X^{\sigma_1}_{r,q}(I)} \le C\big(\|g_1\|_{Y^\sigma_q(I)} + \|g_2\|_{X^\sigma_{r,q}(I)} + \|\nabla_{x,y}\phi\|_{L^2_z(I;C^{-\delta})}\big).$$

On the other hand, for $y \in [-\frac12, 0]$,

$$\phi(x,y) = e^{y|D|}f + \int_0^y e^{-(y'-y)|D|}w(y')dy',$$

from which and (4.6), we infer that

$$\begin{aligned}\|\nabla_{x,y}\phi\|_{X^{\sigma_1}_{r,q}([-\frac12,0])} \le& C\big(\|f\|_{B^{\sigma+1}_{\infty,q}} + \|w\|_{X^{\sigma_1}_{r,q}(I)}\big)\\ \le& C\big(\|f\|_{B^{\sigma+1}_{\infty,q}} + \|g_1\|_{Y^\sigma_q(I)} + \|g_2\|_{X^\sigma_{r,q}(I)} + \|\nabla_{x,y}\phi\|_{L^2_z(I;C^{-\delta})}\big).\end{aligned}$$

Similarly, we can deduce that

$$\|\nabla_{x,y}\phi\|_{X^{\sigma_1}_{r,q}([-1,-\frac12])} \le C\big(\|f\|_{B^{\sigma+1}_{\infty,q}} + \|g_1\|_{Y^\sigma_q(I)} + \|g_2\|_{X^\sigma_{r,q}(I)} + \|\nabla_{x,y}\phi\|_{L^2_z(I;C^{-\delta})}\big).$$

Then the proposition follows by using a bootstrap argument. □

4.2. Flattening the boundary and paralinearization

Motivated by [**3, 24**], we flatten the boundary of $\mathcal{S}$ by a regularized mapping:

$$(x,z) \in \mathbf{R}^d \times I \longmapsto (x, \rho_\delta(x,z)) \in \mathcal{S},$$

where $I = [-1,0]$ and ρ_δ with $\delta > 0$ is given by

$$\rho_\delta(x,z) = z + (1+z)e^{\delta z|D|}\eta(x). \tag{4.9}$$

In the sequel, we denote $\overline{\mathcal{S}} = \mathbf{R}^d \times I$ with $I = [-1,0]$. For a function $f(x,y)$ defined in $\mathcal{S}$, we denote $\widetilde{f}(x,z) = f(x,\rho_\delta(x,z))$.

For any $z\in[-1,0]$, we have

$$\partial_z\rho_\delta = 1+\eta(x)+\big(e^{\delta z|D|}-1\big)\eta(x)+(1+z)\delta e^{\delta z|D|}|D|\eta(x).$$

It is easy to show that

$$\begin{aligned}&\|\big(e^{\delta z|D|}-1\big)\eta+\delta(1+z)e^{\delta z|D|}|D|\eta\|_{L^\infty}\\ &\le \delta\int_z^0\|e^{\delta z'|D|}|D|\eta\|_{L^\infty}dz'+\delta\|e^{\delta z|D|}|D|\eta(x)\|_{L^\infty}\\ &\le C\delta\||D|\eta\|_{L^\infty}\le C\delta\|\eta\|_{C^{1+\varepsilon}}.\end{aligned}$$

Hence, we can take δ small enough depending only on $\|\eta\|_{C^{1+\varepsilon}}$ and h_0 so that

$$\partial_z\rho_\delta(x,z)\ge\frac{h_0}{2}\quad\text{for } (x,z)\in\overline{\mathcal{S}}. \tag{4.10}$$

We have the following regularity information for the regularized map ρ_δ, which can be easily verified by the definition of ρ_δ.

LEMMA 4.3. *For any $\sigma\ge0$, there holds*

$$\|(\nabla\rho_\delta,\partial_z\rho_\delta-1)\|_{X^{\sigma-\frac12}(I)}+\|(\nabla\rho_\delta,\partial_z\rho_\delta-1)\|_{H^\sigma(\mathcal{S})}\le K_\eta\|\eta\|_{H^{\sigma+\frac12}}.$$

Let $p,q\in[1,\infty]$ and $\sigma\in\mathbf{R}$. Then we have

$$\|(\nabla\rho_\delta,\partial_z\rho_\delta-1)\|_{\widetilde{L}^p_z(I;B^\sigma_{\infty,q})}\le K_\eta\|\eta\|_{B^{\sigma+1-\frac1p}_{\infty,q}}.$$

Set $v(x,z)=\phi(x,\rho_\delta(x,z))$. It is easy to find that v satisfies

$$\partial_z^2v+\alpha\Delta v+\beta\cdot\nabla\partial_zv-\gamma\partial_zv=F_0, \tag{4.11}$$

where $F_0=\alpha g(x,\rho_\delta(x,z))\triangleq F_{00}+\nabla_zG_0$ and the coefficients α,β,γ are defined by

$$\alpha=\frac{(\partial_z\rho_\delta)^2}{1+|\nabla\rho_\delta|^2},\beta=-2\frac{\partial_z\rho_\delta\nabla\rho_\delta}{1+|\nabla\rho_\delta|^2},\gamma=\frac{1}{\partial_z\rho_\delta}(\partial_z^2\rho_\delta+\alpha\Delta\rho_\delta+\beta\cdot\nabla\partial_z\rho_\delta). \tag{4.12}$$

We collect the following Sobolev and Hölder regularity information for the elliptic coefficients, which can be proved by using Bony's decomposition (2.3), Lemma 2.16-Lemma 2.18 and Lemma 4.3(see Lemma 4.4 in [**33**]).

LEMMA 4.4. *If $\sigma>\frac32$, then it holds that*

$$\begin{aligned}&\|\alpha-1\|_{X^{\sigma-\frac12}(I)}+\|\beta\|_{X^{\sigma-\frac12}(I)}+\|\gamma\|_{X^{\sigma-\frac32}(I)}\le K_\eta\|\eta\|_{H^{\sigma+\frac12}},\\ &\|\alpha-1\|_{\widetilde{L}^1_z(I;H^{\sigma+\frac12})}+\|\beta\|_{\widetilde{L}^1_z(I;H^{\sigma+\frac12})}+\|\gamma\|_{\widetilde{L}^1_z(I;H^{\sigma-\frac12})}\le K_\eta\|\eta\|_{H^{\sigma+\frac12}},\\ &\|\alpha\|_{\widetilde{L}^\infty_z(I;C^{\frac12+\varepsilon})}+\|\beta\|_{\widetilde{L}^\infty_z(I;C^{\frac12+\varepsilon})}+\|\gamma\|_{\widetilde{L}^\infty_z(I;C^{-\frac12+\varepsilon})}\le K_\eta,\\ &\|\alpha\|_{\widetilde{L}^2_z(I;C^{1+\varepsilon})}+\|\beta\|_{\widetilde{L}^2_z(I;C^{1+\varepsilon})}+\|\gamma\|_{\widetilde{L}^2_z(I;C^{\varepsilon})}\le K_\eta.\end{aligned}$$

LEMMA 4.5. *It holds that*

$$\begin{aligned}&\|(\alpha-1,\beta)\|_{H^\sigma(\mathcal{S})}\le K_\eta\|\eta\|_{H^{\sigma+\frac12}}\quad \textit{for } \sigma\ge0,\\ &\|\gamma\|_{H^{\sigma-1}(\mathcal{S})}\le K_\eta\|\eta\|_{H^{\sigma+\frac12}}\quad \textit{for } \sigma\ge1.\end{aligned}$$

PROOF. The lemma follows from Lemma 2.13 and Lemma 4.3. ∎

In order to obtain the tame elliptic estimates, we paralinearize the elliptic equation (4.11) as

$$\partial_z^2 v + T_\alpha \Delta v + T_\beta \cdot \nabla \partial_z v = F_0 + F_1 + F_2, \tag{4.13}$$

with F_1, F_2 given by

$$F_1 = \gamma \partial_z v, \quad F_2 = (T_\alpha - \alpha)\Delta v + (T_\beta - \beta) \cdot \nabla \partial_z v.$$

Following [**3**], we decouple the equation (4.13) into a forward and a backward parabolic evolution equations:

$$(\partial_z - T_a)(\partial_z - T_A)v = F_0 + F_1 + F_2 + F_3 \triangleq F, \tag{4.14}$$

where

$$\begin{aligned} a &= \frac{1}{2}\big(- i\beta \cdot \xi - \sqrt{4\alpha|\xi|^2 - (\beta \cdot \xi)^2}\big), \\ A &= \frac{1}{2}\big(- i\beta \cdot \xi + \sqrt{4\alpha|\xi|^2 - (\beta \cdot \xi)^2}\big), \\ F_3 &= (T_a T_A - T_\alpha \Delta)v - (T_a + T_A + T_\beta \cdot \nabla)\partial_z v - T_{\partial_z A} v. \end{aligned}$$

The symbols a, A satisfy

Lemma 4.6. *It holds that*

$$\begin{aligned} a(z; x, \xi) \cdot A(z; x, \xi) &= -\alpha(x, z)|\xi|^2, \\ a(z; x, \xi) + A(z; x, \xi) &= -i\beta(x, z) \cdot \xi. \end{aligned}$$

Moreover, $a, A \in \mathcal{M}^1_{\frac{1}{2}+\varepsilon}$ *with the bounds*

$$\mathcal{M}^1_{\frac{1}{2}+\varepsilon}(a) \le K_\eta, \quad \mathcal{M}^1_{\frac{1}{2}+\varepsilon}(A) \le K_\eta.$$

Proof. The two equalities are obvious. Note that

$$4\alpha|\xi|^2 - (\beta \cdot \xi)^2 \ge c_2 |\xi|^2$$

for some $c_2 > 0$ depending only on $\|\eta\|_{C^{\frac{3}{2}+\varepsilon}}$. Then the second statement follows from Lemma 2.12 and Lemma 4.4. □

4.3. Elliptic estimates in Sobolev space

We first present the trace estimate in terms of the H^1 norm of the solution.

Lemma 4.7. *Assume that* $v \in H^1(\overline{\mathcal{S}})$ *is a solution of (*4.11*) in* $\overline{\mathcal{S}}$*. Then we have*

$$\begin{aligned} \|\nabla v\|_{X^{-\frac{1}{2}}(I)} &\le C\|v\|_{H^1(\overline{\mathcal{S}})}, \\ \|\partial_z v\|_{X^{-\frac{1}{2}}(I)} &\le K_\eta\big(\|F_0\|_{Y^{-\frac{1}{2}}(I)} + \|v\|_{H^1(\overline{\mathcal{S}})}\big). \end{aligned}$$

Proof. Let $v_j = \Delta_j v$. Let $\chi(z)$ be a smooth function supported in $[-\frac{3}{4}, 0]$ and $\chi(z) = 1$ on $[-\frac{1}{2}, 0]$. For any $z \in [-\frac{1}{2}, 0]$, we have

$$\begin{aligned} \big\langle \nabla v_j(z), \nabla v_j(z) \big\rangle_{H^{-\frac{1}{2}}} \le& 2\int_{-1}^{z} \chi(z')\big\langle \partial_{z'} \nabla v_j(z'), \nabla v_j(z')\big\rangle_{H^{-\frac{1}{2}}} dz' \\ &+ \int_{-1}^{z} \chi'(z')\big\langle \nabla v_j(z'), \nabla v_j(z')\big\rangle_{H^{-\frac{1}{2}}} dz' \\ \le& C\|\nabla_{x,z} v_j\|^2_{L^2(\overline{\mathcal{S}})} + C\|v\|^2_{L^2(\overline{\mathcal{S}})} \end{aligned}$$

which implies that

$$\|\nabla v\|_{\widetilde{L}^\infty(-\frac12,0;H^{-\frac12})} \le C\|v\|_{H^1(\overline{\mathcal{S}})}.$$

By the equation (4.11) and Lemma 4.8 in [**33**], we get

$$\begin{aligned}
\big\langle \partial_z v_j(z), \partial_z v_j(z)\big\rangle_{H^{-\frac12}} =& 2\int_{-1}^{z} \chi(z')\big\langle \partial_{z'}^2 v_j(z'), \partial_z v_j(z')\big\rangle_{H^{-\frac12}} dz' \\
&+ \int_{-1}^{z} \chi'(z')\big\langle \partial_{z'} v_j(z'), \partial_z v_j(z')\big\rangle_{H^{-\frac12}} dz' \\
=& 2\int_{-1}^{z} \chi(z')\big\langle \Delta_j(F_0 - \alpha\Delta v + \beta\nabla\partial_z v - \gamma\partial_z v), \partial_z v_j\big\rangle_{H^{-\frac12}} dz' \\
&+ \int_{-1}^{z} \chi'(z')\big\langle \partial_{z'} v_j(z'), \partial_z v_j(z')\big\rangle_{H^{-\frac12}} dz' \\
\le& Cc_j\Big(\|F_0\|_{Y^{-\frac12}(I)} + \|\mathrm{div}_x(\alpha\nabla v + \beta\partial_z v)\|_{L^2_z(I;H^{-1})} \\
&+ \|\nabla\alpha\nabla v + \nabla\beta\partial_z v + \gamma\partial_z v\|_{\widetilde{L}^1_z(I;H^{-\frac12})} \\
&+ \|\nabla_{x,z} v\|_{L^2(I\times\mathbf{R}^d)}\Big)\|\partial_z v\|_{X^{-\frac12}(I)} \\
\le& K_\eta c_j\big(\|F_0\|_{Y^{-\frac12}(I)} + \|v\|_{H^1(\overline{\mathcal{S}})}\big)\|\partial_z v\|_{X^{-\frac12}(I)},
\end{aligned}$$

with $\|\{c_j\}\|_{\ell^1} \le 1$, which implies that

$$\|\partial_z v\|^2_{\widetilde{L}^\infty(-\frac12,0;H^{-\frac12})} \le K_\eta\big(\|F_0\|_{Y^{-\frac12}(I)} + \|\nabla_{x,z} v\|_{L^2(\overline{\mathcal{S}})}\big)\|\nabla_{x,z} v\|_{X^{-\frac12}(I)}.$$

The same estimates hold for $z \in [-1, -\frac12]$ by a similar cut-off argument. This gives

$$\|\nabla_{x,z} v\|^2_{X^{-\frac12}(I)} \le K_\eta\big(\|F_0\|_{Y^{-\frac12}(I)} + \|\nabla_{x,z} v\|_{L^2(\overline{\mathcal{S}})}\big)\|\nabla_{x,z} v\|_{X^{-\frac12}(I)},$$

which implies the desired result. □

REMARK 4.8. If $\eta \in H^s(\mathbf{R}^d)$ for $s > \frac{d}{2} + 1$, then we have

$$\|\nabla_{x,z} v\|_{X^{-\frac12}(I)} \le C(\|\eta\|_{H^s}, h_0)\big(\|F_0\|_{Y^{-\frac12}(I)} + \|v\|_{H^1(\overline{\mathcal{S}})}\big).$$

Indeed, we have $\nabla\alpha, \nabla\beta, \gamma \in H^{s-\frac32}(\overline{\mathcal{S}})$, which implies by Lemma 2.10 that

$$\|\nabla\alpha\nabla v + \nabla\beta\partial_z v + \gamma\partial_z v\|_{L^1_z(I;H^{-\frac12})} \le C(\|\eta\|_{H^s}, h_0)\|\nabla_{x,z} v\|_{L^2(\overline{\mathcal{S}})}.$$

The following elliptic estimates will be used to estimate the velocity in the proof of well-posedness. Here and in what follows, we denote by $P_{\sigma,\eta} = C\big(\|\eta\|_{H^\sigma}\big)$ an increasing function depending on h_0, which may be different from line to line.

Let us first recall the following elliptic estimate for tangential derivatives, which has been essentially proved in [**3**].

PROPOSITION 4.9. *Let $v \in H^1(\overline{\mathcal{S}})$ be a solution of (4.11) in $\overline{\mathcal{S}}$ with $v(x,0) = f(x)$ and $v(x,-1) = f_b(x)$. Assume that $\eta \in H^s(\mathbf{R}^d)$ for $s > \frac{d}{2} + \frac32$. Then for any $\sigma \in [-\frac12, s - \frac12]$, it holds that*

$$\|\nabla_{x,z} v\|_{X^{\sigma-\frac12}(I)} \le P_{s,\eta}\big(\|\nabla_{x,z} v\|_{L^2(\overline{\mathcal{S}})} + \|(f, f_b)\|_{H^{\sigma+\frac12}} + \|F_0\|_{L^2_z(I:H^{\sigma-1})}\big),$$

$$\|\nabla_{x,z} v\|_{X^{\sigma-\frac12}([a,0])} \le P_{s,\eta}\big(\|\nabla_{x,z} v\|_{L^2(\overline{\mathcal{S}})} + \|f\|_{H^{\sigma+\frac12}} + \|F_0\|_{L^2_z(I:H^{\sigma-1})}\big),$$

for any $a \in (-1, 0)$.

Now we present the elliptic estimate for full derivatives.

PROPOSITION 4.10. *Let $v \in H^1(\overline{\mathcal{S}})$ be a solution of (4.11) in $\overline{\mathcal{S}}$ with $v(x,0) = f(x)$ and $v(x,-1) = f_b(x)$. Assume that $\eta \in H^s(\mathbf{R}^d)$ for $s > \frac{d}{2} + \frac{3}{2}$. Then it holds that*

$$\|\nabla_{x,z} v\|_{H^{s-\frac{1}{2}}(\overline{\mathcal{S}})} \le P_{s,\eta}\big(\|\nabla_{x,z} v\|_{L^2(\overline{\mathcal{S}})} + \|(f, f_b)\|_{H^s} + \|F_0\|_{H^{s-\frac{3}{2}}(\overline{\mathcal{S}})}\big).$$

We need the following lemma.

LEMMA 4.11. *Let $\sigma = s - \frac{1}{2}$ and $\sigma_1 = \sigma - [\sigma]$. Then for any positive integer $k \le [\sigma]$, there holds*

$$\begin{aligned}&\|\partial_z^{k-1}(\alpha\Delta v + \beta\cdot\nabla\partial_z v - \gamma\partial_z v)\|_{L^2_z(I;H^{\sigma_1})}\\ &\le P_{s,\eta}\Big(\|\langle D\rangle^{\sigma_1}\nabla v\|_{H^k(\overline{\mathcal{S}})} + \|\langle D\rangle^{\sigma_1+\frac{1}{2}}\partial_z v\|_{H^{k-1}(\overline{\mathcal{S}})} + \|\nabla_{x,z} v\|_{L^\infty(\overline{\mathcal{S}})}\\ &\qquad + \|\langle D\rangle^{\sigma_1+\frac{1}{2}}\nabla_{x,z} v\|_{L^\infty_z(I;L^2)} + \|\langle D\rangle^{\frac{d}{2}}\nabla_{x,z} v\|_{L^\infty_z(I;L^2)}\Big).\end{aligned}$$

PROOF. For $k = 1$, we have by Lemma 2.11 and Lemma 4.4 that

$$\begin{aligned}&\|\alpha\Delta v + \beta\cdot\nabla\partial_z v - \gamma\partial_z v\|_{L^2_z(I;H^{\sigma_1})}\\ &\le C\big(\|(\alpha-1,\beta)\|_{L^\infty} + 1\big)\|\langle D\rangle^{\sigma_1}\nabla v\|_{H^1(\overline{\mathcal{S}})} + \|\gamma\|_{L^\infty_z(I;C^{-\frac{1}{2}})}\|\langle D\rangle^{\sigma_1+\frac{1}{2}}\partial_z v\|_{L^2(\overline{\mathcal{S}})}\\ &\qquad + C\|\nabla_{x,z} v\|_{L^\infty(\overline{\mathcal{S}})}\big(\|(\alpha-1,\beta)\|_{L^2(I;H^{1+\sigma_1})} + \|\gamma\|_{L^2(I;H^{\sigma_1})}\big)\\ &\le P_{s,\eta}\big(\|\nabla_{x,z} v\|_{L^\infty(\overline{\mathcal{S}})} + \|\langle D\rangle^{\sigma_1}\nabla v\|_{H^1(\overline{\mathcal{S}})} + \|\langle D\rangle^{\sigma_1+\frac{1}{2}}\partial_z v\|_{L^2(\overline{\mathcal{S}})}\big).\end{aligned}$$

Here we used $\sigma_1 + \frac{3}{2} \le s$ and $s > \frac{d}{2} + \frac{3}{2}$.

Next we consider the case of $k \ge 2$. Let $(\overline{v}, \overline{\alpha}, \overline{\beta}, \overline{\gamma})$ be an extension of $(v, \alpha, \beta, \gamma)$ to $\mathbf{R}^{d+1}$ so that

$$\begin{aligned}&\|\nabla_{x,z}\overline{v}\|_{H^\sigma(\mathbf{R}^{d+1})} \le C\|\nabla_{x,z} v\|_{H^\sigma(\overline{\mathcal{S}})},\\ &\|(\overline{\alpha}-1,\overline{\beta})\|_{H^\sigma(\mathbf{R}^{d+1})} \le C\|(\alpha-1,\beta)\|_{H^\sigma(\overline{\mathcal{S}})},\\ &\|\overline{\gamma}\|_{H^{\sigma-1}(\mathbf{R}^{d+1})} \le C\|\gamma\|_{H^{\sigma-1}(\overline{\mathcal{S}})}.\end{aligned}$$

Using Bony's decomposition (2.3), we write

$$\begin{aligned}\overline{\beta}\nabla\partial_z\overline{v} &= T_{\overline{\beta}}\nabla\partial_z\overline{v} + T_{\nabla\partial_z\overline{v}}\overline{\beta} + R(\overline{\beta}, \nabla\partial_z\overline{v})\\ &= T_{\overline{\beta}}\nabla\partial_z\overline{v} + \partial_z T_{\nabla\overline{v}}\overline{\beta} - T_{\nabla\overline{v}}\partial_z\overline{\beta} + R(\overline{\beta}, \nabla\partial_z\overline{v}),\end{aligned}$$

from which and Lemma 2.14, we infer that for any $\epsilon > 0$

$$\begin{aligned}\|\overline{\beta}\nabla\partial_z\overline{v}\|_{H^{k-1,\sigma_1}} &\le C\big(\|\overline{\beta}\|_{L^\infty} + \|\langle D\rangle^{\sigma_1+\epsilon}\overline{\beta}\|_{L^\infty_z(\mathbf{R};L^2)} + \|\langle D\rangle^{\frac{d}{2}}\overline{\beta}\|_{L^\infty_z(\mathbf{R};L^2)}\big)\|\nabla\overline{v}\|_{H^{k,\sigma_1}}\\ &\quad + C\big(\|\nabla_{x,z}\overline{v}\|_{L^\infty} + \|\langle D\rangle^{\sigma_1+\epsilon}\nabla\overline{v}\|_{L^\infty_z(\mathbf{R};L^2)}\\ &\quad + \|\langle D\rangle^{\frac{d}{2}}\nabla\overline{v}\|_{L^\infty_z(\mathbf{R};L^2)}\big)\|\overline{\beta}\|_{H^{k,\sigma_1}}.\end{aligned}$$

Due to $\sigma_1 + \epsilon + 1 \le s$ for some $\epsilon \in (0,1]$ and $s > \frac{d}{2} + \frac{3}{2}$, it follows from Lemma 4.5 that

$$\|\overline{\beta}\|_{L^\infty} + \|\langle D\rangle^{\sigma_1+\epsilon}\overline{\beta}\|_{L^\infty_z(\mathbf{R};L^2)} + \|\langle D\rangle^{\frac{d}{2}}\overline{\beta}\|_{L^\infty_z(\mathbf{R};L^2)} + \|\overline{\beta}\|_{H^{k,\sigma_1}} \le P_{s,\eta}.$$

This shows that

$$\begin{aligned}\|\langle D\rangle^{\sigma_1}(\beta\nabla\partial_z v)\|_{H^{k-1}(\overline{\mathcal{S}})} &\le P_{s,\eta}\big(\|\langle D\rangle^{\sigma_1}\nabla v\|_{H^k(\overline{\mathcal{S}})} + \|\nabla_{x,z} v\|_{L^\infty(\overline{\mathcal{S}})}\\ &\qquad + \|\langle D\rangle^{\sigma_1+\epsilon}\nabla v\|_{L^\infty_z(I;L^2)} + \|\langle D\rangle^{\frac{d}{2}}\nabla v\|_{L^\infty_z(I;L^2)}\big).\end{aligned}$$

Similarly, we have

$$\|\langle D\rangle^{\sigma_1}(\alpha\Delta v)\|_{H^{k-1}(\overline{\mathcal{S}})} \leq P_{s,\eta}\big(\|\langle D\rangle^{\sigma_1}\nabla v\|_{H^k(\overline{\mathcal{S}})} + \|\nabla_{x,z}v\|_{L^\infty(\overline{\mathcal{S}})} + \|\langle D\rangle^{\sigma_1+\epsilon}\nabla v\|_{L^\infty_z(I;L^2)} + \|\langle D\rangle^{\frac{d}{2}}\nabla v\|_{L^\infty_z(I;L^2)}\big).$$

While, we can see from the proof of Lemma 2.14 that

$$\begin{aligned}\|\langle D\rangle^{\sigma_1}\overline{\gamma}\partial_z\overline{v}\|_{H^{k-1,\sigma_1}} \leq{}& C\big(\|\overline{\gamma}\|_{L^\infty_z(\mathbf{R};C^{-\frac12})} + \|\langle D\rangle^{\sigma_1+\epsilon}\overline{\gamma}\|_{L^2} \\ &+ \|\langle D\rangle^{\frac{d}{2}}\overline{\gamma}\|_{L^2}\big)\|\langle D\rangle^{\sigma_1+\frac12}\partial_z\overline{v}\|_{H^{k-1}} \\ &+ C\big(\|\nabla_{x,z}\overline{v}\|_{L^\infty} + \|\langle D\rangle^{\sigma_1+\epsilon}\partial_z\overline{v}\|_{L^\infty_y(\mathbf{R};L^2)} \\ &+ \|\langle D\rangle^{\frac{d}{2}}\partial_z\overline{v}\|_{L^\infty_y(\mathbf{R};L^2)}\big)\|\overline{\gamma}\|_{H^{k-1,\sigma_1}}.\end{aligned}$$

We know from Lemma 4.4 and Lemma 4.5 that

$$\|\overline{\gamma}\|_{L^\infty_z(\mathbf{R};C^{-\frac12})} + \|\langle D\rangle^{\sigma_1+\epsilon}\overline{\gamma}\|_{L^2} + \|\langle D\rangle^{\frac{d}{2}}\overline{\gamma}\|_{L^2} + \|\overline{\gamma}\|_{H^{k-1,\sigma_1}} \leq P_{s,\eta}.$$

Then we infer that

$$\|\gamma\partial_z v\|_{H^{k-1}(\overline{\mathcal{S}})} \leq P_{s,\eta}\big(\|\langle D\rangle^{\sigma_1}\partial_z v\|_{H^{k-\frac12}(\overline{\mathcal{S}})} + \|\nabla_{x,z}v\|_{L^\infty(\overline{\mathcal{S}})} + \|\langle D\rangle^{\sigma_1+\epsilon}\partial_z v\|_{L^\infty_z(I;L^2)} + \|\langle D\rangle^{\frac{d}{2}}\partial_z v\|_{L^\infty_z(I;L^2)}\big).$$

Summing up the above estimates, we conclude the lemma. □

PROOF OF PROPOSITION 4.10. Let $\sigma = s - \frac12$. Proposition 4.9 ensures that

$$\begin{aligned}&\|\nabla_{x,z}v\|_{L^\infty_z(I;H^{\sigma-\frac12})} + \|\nabla_{x,z}v\|_{L^2_z(I;H^\sigma)} \\ &\leq P_{s,\eta}\big(\|\nabla_{x,z}v\|_{L^2(\overline{\mathcal{S}})} + \|(f,f_b)\|_{H^{\sigma+\frac12}} + \|F_0\|_{L^2_z(I:H^{\sigma-1})}\big).\end{aligned}\tag{4.15}$$

Assume that for $1 \leq \ell \leq [\sigma]-1$, there holds

$$\|\nabla^{\ell+1}_{x,z}v\|_{L^2_z(I;H^{\sigma-\ell})} \leq P_{s,\eta}\big(\|\nabla_{x,z}v\|_{L^2(\overline{\mathcal{S}})} + \|(f,f_b)\|_{H^{\sigma+\frac12}} + \|F_0\|_{H^{\sigma-1}(\overline{\mathcal{S}})}\big).\tag{4.16}$$

We will prove that the inequality holds for $\ell = [\sigma] \triangleq k$. Let $\sigma_1 = \sigma - k$. Using the equation (4.11) and Lemma 4.11, we deduce that

$$\begin{aligned}&\|\partial_z^{k+1}v\|_{L^2_z(I;H^{\sigma_1})} \leq \|\partial_z^{k-1}(\alpha\Delta v + \beta\cdot\nabla\partial_z v - \gamma\partial_z v)\|_{L^2_z(I;H^{\sigma_1})} + \|\partial_z^{k-1}F_0\|_{L^2_z(I;H^{\sigma_1})} \\ &\leq P_{s,\eta}\Big(\|\langle D\rangle^{\sigma_1}\nabla v\|_{H^k(\overline{\mathcal{S}})} + \|\langle D\rangle^{\sigma_1+\frac12}\partial_z v\|_{H^{k-1}(\overline{\mathcal{S}})} + \|\nabla_{x,z}v\|_{L^\infty_z(I;H^{\sigma-\frac12})} \\ &\quad + \|\langle D\rangle^{\sigma_1+\frac12}\nabla_{x,z}v\|_{L^\infty_z(I;L^2)} + \|\langle D\rangle^{\frac{d}{2}}\nabla_{x,z}v\|_{L^\infty_z(I;L^2)}\Big) + \|F_0\|_{H^{\sigma-1}(\overline{\mathcal{S}})}.\end{aligned}\tag{4.17}$$

Note that $\sigma_1 + \frac12 \leq \sigma - \frac12, \frac{d}{2} < \sigma - \frac12$. This together with (4.15) and (4.16) ensures that the inequality (4.16) holds for $\ell = [\sigma]$.

Using Lemma 2.13, Lemma 4.5 and the interpolation, we obtain

$$\begin{aligned}\|\partial_z^{k+1}v\|_{H^{\sigma_1}(\overline{\mathcal{S}})} \leq{}& \|\partial_z^{k-1}(\alpha\Delta v + \beta\cdot\nabla\partial_z v - \gamma\partial_z v)\|_{H^{\sigma_1}(\overline{\mathcal{S}})} + \|\partial_z^{k-1}F_0\|_{H^{\sigma_1}(\overline{\mathcal{S}})} \\ \leq{}& P_{s,\eta}\big(\|\nabla\nabla_{x,z}v\|_{H^{k-1+\sigma_1}(\overline{\mathcal{S}})} + \|\langle D\rangle^{\frac12}\partial_z v\|_{H^{k-1+\sigma_1}(\overline{\mathcal{S}})}\big) + \|F_0\|_{H^{\sigma-1}(\overline{\mathcal{S}})} \\ \leq{}& P_{s,\eta}\big(\|\nabla_{x,z}v\|_{L^2_z(I;H^{k+\sigma_1})} + \|\partial_z^{k+1}v\|_{L^2_z(I;H^{\sigma_1})}\big) + \|F_0\|_{H^{\sigma-1}(\overline{\mathcal{S}})},\end{aligned}$$

which together with (4.15) and (4.17) implies the proposition. □

4.4. Tame elliptic estimates

To prove the break-down criterion, we need to establish the tame elliptic estimates. In this subsection, we assume that $\eta \in H^{s+\frac{1}{2}}(\mathbf{R}^d)$ for $s > \frac{d}{2} + 1$.

The following elliptic estimate will be used to estimate the Drichlet-Neumann operator and the pressure.

PROPOSITION 4.12. *Let $I_0 = [a, 0]$ for $a \in (-1, 0)$. Assume that $v \in H^{s+1}(\overline{\mathcal{S}})$ is a solution of (4.11) in $\overline{\mathcal{S}}$ with $v(x, 0) = f(x)$. Then there exists an interval $I_1 \subset (-1, 0)$ so that for all $\sigma \in [-\frac{1}{2}, s - \frac{1}{2}]$,*

$$\begin{aligned}\|\nabla_{x,z}v\|_{X^\sigma(I_0)} \leq &K_\eta\big(\|\nabla_{x,z}v\|_{L^2(\overline{\mathcal{S}})} + \|f\|_{H^{\sigma+1}} + \|F_{00}\|_{Y^\sigma(I)} \\ &+ \|G_0\|_{X^\sigma(I)} + \|\eta\|_{H^{s+\frac{1}{2}}}\|\nabla_{x,z}v\|_{L^\infty(\widetilde{\mathcal{S}})}\big).\end{aligned}$$

Here $\widetilde{\mathcal{S}} = \mathbf{R}^d \times I_1$. Moreover, if $\sigma < s - \frac{1}{2}$, $\|\nabla_{x,z}v\|_{L^\infty(\widetilde{\mathcal{S}})}$ can be replaced by $\|\nabla_{x,z}v\|_{L^\infty_z(I_1;C^0)}$.

The proof of the proposition needs the following lemma, which can be proved by using Bony's decomposition, Lemma 2.16 and Lemma 2.18. Since the proof is almost the same as those in [**33**], we omit the details.

LEMMA 4.13. *Let $I_1 \subseteq I$ be an interval and $\widetilde{\mathcal{S}} = \mathbf{R}^d \times I_1$. For any $\sigma \leq s - \frac{1}{2}$, it holds that*

$$\begin{aligned}&\|F_1\|_{Y^\sigma(I_1)} \leq K_\eta\big(\|\partial_z v\|_{L^2_z(I_1;H^\sigma)} + \|\partial_z v\|_{L^\infty(\widetilde{\mathcal{S}})}\|\eta\|_{H^{s+\frac{1}{2}}}\big),\\ &\|F_2\|_{Y^\sigma(I_1)} \leq K_\eta\|\nabla_{x,z}v\|_{L^\infty(\widetilde{\mathcal{S}})}\|\eta\|_{H^{s+\frac{1}{2}}},\\ &\|F_3\|_{Y^\sigma(I_1)} \leq K_\eta\|\nabla v\|_{L^2_z(I_1;H^\sigma)} \quad \textit{for any} \quad \sigma \in \mathbf{R}.\end{aligned}$$

If $\sigma < s - \frac{1}{2}$, $\|\nabla_{x,z}v\|_{L^\infty(\widetilde{\mathcal{S}})}$ can be replaced by $\|\nabla_{x,z}v\|_{L^\infty_z(I_1;C^0)}$.

PROOF OF PROPOSITION 4.12. As in [**3**], the proof uses the induction argument. Suppose that there exits $I_2 = [\xi_0, 0] \subseteq I_1$ with $\xi_0 \in (-1, a)$ such that for some $\sigma \in [-\frac{1}{2}, s - \frac{1}{2})$, there holds

$$\begin{aligned}\|\nabla_{x,z}v\|_{X^\sigma(I_2)} \leq &K_\eta\big(\|\nabla_{x,z}v\|_{L^2(\overline{\mathcal{S}})} + \|f\|_{H^{\sigma+1}} + \|F_{00}\|_{Y^\sigma(I)} \\ &+ \|G_0\|_{X^\sigma(I)} + \|\eta\|_{H^{s+\frac{1}{2}}}\|\nabla_{x,z}v\|_{L^\infty(\widetilde{\mathcal{S}})}\big).\end{aligned}$$

This is indeed true for $\sigma = -\frac{1}{2}$ by Lemma 4.7. We will show that the above inequality still holds when σ and I_2 are replaced by $\sigma+\delta_1$ and I_3, where $\sigma+\delta_1 \leq s-\frac{1}{2}$ with $\delta_1 \in (0, \frac{1}{2}]$, and $I_3 = [\xi_1, 0]$ with $\xi_1 = \xi_0 + \frac{1}{2}(a - \xi_0) \in (\xi_0, a)$.

Let χ be a smooth function such that $\chi(\xi_0) = 0$ and $\chi(z) = 1$ for $z \geq \xi_1$. Set $w = \chi(z)(\partial_z - T_A)v - \chi(z)G_0$. Then w satisfies

$$\partial_z w - T_a w = F', \quad w(\xi_0) = 0,$$

where $F' = \chi(z)\big(F_{00} + F_1 + F_2 + F_3\big) + \chi'(z)((\partial_z - T_A)v - G_0) + \chi(z)T_a G_0$.

We deduce from Proposition 2.6 and Lemma 4.6 that for $\delta_1 \leq \frac{1}{2}$,

$$\|(\partial_z - T_A)v\|_{L^2(I_2;H^{\sigma+\delta_1})} \leq K_\eta\|\nabla_{x,z}v\|_{L^2(I_2;H^{\sigma+\delta_1})},$$

which along with Lemma 4.13 gives

$$\begin{aligned}\|F'\|_{Y^{\sigma+\delta_1}(I_2)} \leq &\|F_0\|_{Y^{\sigma+\delta_1}(I_2)} + K_\eta\|G_0\|_{X^\sigma(I_2)} \\ &+ K_\eta\big(\|\nabla_{x,z}v\|_{L^\infty(I_2\times\mathbf{R}^d)}\|\eta\|_{H^{s+\frac{1}{2}}} + \|\nabla_{x,z}v\|_{X^\sigma(I_2)}\big).\end{aligned} \tag{4.18}$$

Then it follows from Proposition 3.1 that

$$\begin{aligned}\|w\|_{X^{\sigma+\delta_1}(I_2)} &\le K_\eta\|F\|_{Y^{\sigma+\delta_1}(I_2)}\\ &\le K_\eta\big(\|F_0\|_{Y^{r+\delta_1}(I)}+\|G_0\|_{X^\sigma(I)}+\|\nabla_{x,z}v\|_{X^\sigma(I_2)}+\|\nabla_{x,z}v\|_{L^\infty(I_2\times\mathbf{R}^d)}\|\eta\|_{H^{s+\frac12}}\big).\end{aligned}\tag{4.19}$$

To obtain the estimate of v, we consider the backward parabolic equation

$$\partial_z v - T_A v = w + G_0 \quad \text{in } I_3, \quad v|_{z=0}=f.$$

Take ∇ to the equation of v to get

$$(\partial_z - T_A)\nabla v = \nabla(w+G_0) + T_{\nabla A}v, \quad \nabla v|_{z=0} = \nabla f.$$

By Remark 2.8 and Lemma 4.6, we have

$$\|T_{\nabla A}v\|_{L^2_z(I_3;H^{\sigma+\delta_1-\frac12})} \le K_\eta\|\nabla v\|_{L^2_z(I_3;H^{\sigma+\delta_1})}.$$

Then Proposition 3.1, (4.19) and the induction assumption ensure that

$$\begin{aligned}\|\nabla v\|_{X^{\sigma+\delta_1}(I_3)} &\le K_\eta\big(\|w\|_{X^{\sigma+\delta_1}(I_3)}+\|G_0\|_{X^{\sigma+\delta_1}(I_3)}\\ &\quad+\|f\|_{H^{\sigma+1+\delta_1}}+\|\nabla v\|_{L^2(I_3;H^{\sigma+\delta_1})}\big)\\ &\le K_\eta\big(\|f\|_{H^{\sigma+1+\delta_1}}+\|F_0\|_{Y^{\sigma+\delta_1}(I)}+\|G_0\|_{X^{\sigma+\delta_1}(I_3)}\\ &\quad+\|\nabla_{x,z}v\|_{X^\sigma(I_2)}+\|\nabla_{x,z}v\|_{L^\infty(I_2\times\mathbf{R}^d)}\|\eta\|_{H^{s+\frac12}}\big)\\ &\le K_\eta\big(\|\nabla_{x,z}v\|_{L^2(\overline{\mathcal{S}})}+\|f\|_{H^{\sigma+\delta_1+1}}+\|F_0\|_{Y^{\sigma+\delta_1}(I)}\\ &\quad+\|G_0\|_{X^{\sigma+\delta_1}(I_3)}+\|\nabla_{x,z}v\|_{L^\infty(I_2\times\mathbf{R}^d)}\|\eta\|_{H^{s+\frac12}}\big).\end{aligned}$$

The same estimate holds for $\partial_z v$ by using $\partial_z v = T_A v + w + G_0$.

In the case when $\sigma+\delta_1 < s-\frac12$, $\|\nabla_{x,z}v\|_{L^\infty(I_2\times\mathbf{R}^d)}$ in (4.18) can be replaced by $\|\nabla_{x,z}v\|_{L^\infty_z(I_2;B^0_{\infty,\infty})}$ by Lemma 4.13, and so does the final result. □

The following elliptic estimate will be used to estimate the velocity in the proof of break-down criterion.

PROPOSITION 4.14. *Assume that $v\in H^{s+\frac12}(\overline{\mathcal{S}})$ is a solution of (4.11) in $\overline{\mathcal{S}}$ with $v(x,0)=f(x)$ and $v(x,-1)=f_b(x)$. Let $k=s-\frac12$ be an integer. Then it holds that*

$$\begin{aligned}\|\nabla_{x,z}v\|_{H^k(\overline{\mathcal{S}})} &\le K_\eta\big(\|\nabla_{x,z}v\|_{L^2(\overline{\mathcal{S}})}+\|(f,f_b)\|_{H^s}\\ &\quad+\|F_0\|_{H^{k-1}(\overline{\mathcal{S}})}+\|\eta\|_{H^{s+\frac12}}\|\nabla_{x,z}v\|_{L^\infty(\overline{\mathcal{S}})}\big).\end{aligned}$$

Let us first present the tame estimate for the tangential derivatives.

LEMMA 4.15. *Assume that $v\in H^{s+1}(\overline{\mathcal{S}})$ is a solution of (4.11) in $\overline{\mathcal{S}}$ with $v(x,0)=f(x)$ and $v(x,-1)=f_b(x)$. Then for all $\sigma\in[-\frac12,s-\frac12]$, it holds that*

$$\|\nabla_{x,z}v\|_{X^\sigma(I)} \le K_\eta\big(\|\nabla_{x,z}v\|_{L^2(\overline{\mathcal{S}})}+\|(f,f_b)\|_{H^{\sigma+1}}+\|F_0\|_{Y^\sigma(I)}+\|\eta\|_{H^{s+\frac12}}\|\nabla_{x,z}v\|_{L^\infty(\overline{\mathcal{S}})}\big).$$

PROOF. Because the proof is similar to Proposition 4.12, we just present a sketch. Assume that for some $\sigma\in[-\frac12,s-\frac12)$, there holds

$$\|\nabla_{x,z}v\|_{X^\sigma(I)} \le K_\eta\big(\|\nabla_{x,z}v\|_{L^2(\overline{\mathcal{S}})}+\|f\|_{H^{\sigma+1}}+\|F_0\|_{Y^\sigma(I)}+\|\eta\|_{H^{s+\frac12}}\|\nabla_{x,z}v\|_{L^\infty(\overline{\mathcal{S}})}\big).$$

We show that the above inequality still holds when σ is replaced by $\sigma+\delta_1$, where $\sigma+\delta_1\le s-\frac12$ with $\delta_1\in(0,\frac12]$.

Let χ be a smooth function such that $\chi(-1)=0$ and $\chi(z)=1$ for $z\in[-\frac{1}{2},0]$. Set $w=\chi(z)(\partial_z-T_A)v$. Then (w,v) satisfies

$$\partial_z w - T_a w = \widetilde{F}, \quad w(-1)=0,$$
$$\partial_z v - T_A v = w, \quad v(0)=f,$$

where $\widetilde{F}=\chi(z)\big(F_0+F_1+F_2+F_3\big)+\chi'(z)(\partial_z-T_A)v$. Using the same argument of Proposition 4.12, we can deduce that

$$\begin{aligned}&\|\nabla_{x,z}v\|_{X^{\sigma+\delta_1}([-\frac{1}{2},0])}\\ &\le K_\eta\big(\|\nabla_{x,z}v\|_{L^2(\overline{\mathcal{S}})}+\|f\|_{H^{\sigma+\delta_1+1}}+\|F_0\|_{Y^{\sigma+\delta_1}(I)}+\|\nabla_{x,z}v\|_{L^\infty(\overline{\mathcal{S}})}\|\eta\|_{H^{s+\frac{1}{2}}}\big).\end{aligned}$$

On the other hand, the equation (4.13) can also be decoupled into a backward and forward parabolic evolution equations, i.e.,

$$(\partial_z-T_A)(\partial_z-T_a)v=F_0+F_1+F_2+\widetilde{F}_3,$$

where

$$\widetilde{F}_3=(T_AT_a-T_\alpha\Delta)v-(T_a+T_A+T_\beta\cdot\nabla)\partial_z v-T_{\partial_z a}v.$$

Now we let χ be a smooth function such that $\chi(0)=0$ and $\chi(z)=1$ for $z\in[-1,-\frac{1}{2}]$. Set $w=\chi(z)(\partial_z-T_a)v$. Then (w,v) satisfies

$$\partial_z w - T_A w = \widetilde{F}, \quad w(0)=0,$$
$$\partial_z v - T_a v = w, \quad v(-1)=f_b,$$

A similar argument ensures that

$$\begin{aligned}&\|\nabla_{x,z}v\|_{X^{\sigma+\delta_1}([-1-\frac{1}{2}])}\\ &\le K_\eta\big(\|\nabla_{x,z}v\|_{L^2(\overline{\mathcal{S}})}+\|f_b\|_{H^{\sigma+\delta_1+1}}+\|F_0\|_{Y^{\sigma+\delta_1}(I)}+\|\nabla_{x,z}v\|_{L^\infty(\overline{\mathcal{S}})}\|\eta\|_{H^{s+\frac{1}{2}}}\big).\end{aligned}$$

This together with the estimates in the interval $[-\frac{1}{2},0]$ gives the lemma. □

PROOF OF PROPOSITION 4.14. Lemma 4.15 ensures that

$$\begin{aligned}\|\nabla_{x,z}v\|_{L^\infty(I;H^{k-\frac{1}{2}})}+\|\nabla_{x,z}v\|_{L^2(I;H^k)}\le K_\eta\big(&\|\nabla_{x,z}v\|_{L^2(\overline{\mathcal{S}})}+\|(f,f_b)\|_{H^s}\\ &+\|F_0\|_{L^2_z(I:H^{k-1})}+\|\eta\|_{H^{s+\frac{1}{2}}}\|\nabla_{x,z}v\|_{L^\infty(\overline{\mathcal{S}})}\big).\end{aligned}\tag{4.20}$$

Let us assume that for $0\le \ell\le k-1$, there holds

$$\begin{aligned}\|\nabla^{\ell+1}_{x,z}v\|_{L^2(I;H^{k-\ell})}\le &K_\eta\Big(\|\nabla_{x,z}v\|_{L^2(\overline{\mathcal{S}})}+\|(f,f_b)\|_{H^s}\\ &+\sum_{\ell'\le \ell-1}\|\nabla^{\ell'}_{x,z}F_0\|_{L^2(\overline{\mathcal{S}})}+\|\eta\|_{H^{s+\frac{1}{2}}}\|\nabla_{x,z}v\|_{L^\infty(\overline{\mathcal{S}})}\Big).\end{aligned}\tag{4.21}$$

Next we show that the inequality holds for $\ell=k$. Using the equation (4.11), we get

$$\|\partial_z^{k+1}v\|_{L^2(\overline{\mathcal{S}})}\le \|\partial_z^{k-1}(\alpha\Delta v+\beta\cdot\nabla\partial_z v-\gamma\partial_z v)\|_{L^2(\overline{\mathcal{S}})}+\|\partial_z^{k-1}F_0\|_{L^2(\overline{\mathcal{S}})}.\tag{4.22}$$

As in the proof of Proposition 4.10, let $(\overline{v},\overline{\alpha},\overline{\beta},\overline{\gamma})$ be an extension of (v,α,β,γ) to $\mathbf{R}^{d+1}$ keeping the corresponding Sobolev norm. We infer from Lemma 2.11 that

$$\begin{aligned}\|(\overline{\alpha}-1)\Delta\overline{v}\|_{H^{k-1}}\le &C\|\overline{\alpha}-1\|_{L^\infty}\|\nabla^2\overline{v}\|_{H^{k-1}}+C\|\nabla\overline{v}\|_{L^\infty}\|\overline{\alpha}-1\|_{H^k}\\ \le &C\|\alpha-1\|_{L^\infty(\overline{\mathcal{S}})}\|\nabla^2 v\|_{H^{k-1}(\overline{\mathcal{S}})}+C\|\nabla v\|_{L^\infty(\overline{\mathcal{S}})}\|\alpha-1\|_{H^k(\overline{\mathcal{S}})}\end{aligned}$$

Similarly, we have

$$\|\overline{\beta}\nabla\partial_z\overline{v}\|_{H^{k-1}} \leq C\|\beta\|_{L^\infty(\overline{\mathcal{S}})}\|\nabla\nabla_{x,z}v\|_{H^{k-1}(\overline{\mathcal{S}})} + C\|\nabla_x v\|_{L^\infty(\overline{\mathcal{S}})}\|\beta\|_{H^k(\overline{\mathcal{S}})}.$$

Using Bony's decomposition (2.3), we write

$$\overline{\gamma}\partial_z\overline{v} = T_{\overline{\gamma}}\partial_z\overline{v} + T_{\partial_z\overline{v}}\overline{\gamma} + R(\partial_z\overline{v},\overline{\gamma}).$$

It follows from Lemma 2.2 that

$$\begin{aligned}\|S_{j-3}\overline{\gamma}\Delta_j(\partial_z\overline{v})\|_{L^2} \leq & C\|\overline{\gamma}\|_{L^\infty_z(\mathbf{R};C^{-\frac12})}\|\langle D\rangle^{\frac12}\Delta_j(\partial_z\overline{v})\|_{L^2}\\ \leq & Cc_j 2^{-j(k-1)}\|\overline{\gamma}\|_{L^\infty_z(\mathbf{R};C^{-\frac12})}\|\langle D\rangle^{\frac12}\partial_z\overline{v}\|_{H^{k-1}}\end{aligned}$$

with $\|\{c_j\}\|_{\ell^2} \leq 1$. Then Lemma 2.4 ensures that

$$\|T_{\overline{\gamma}}\partial_z\overline{v}\|_{H^{k-1}} \leq C\|\overline{\gamma}\|_{L^\infty_z(\mathbf{R};C^{-\frac12})}\|\langle D\rangle^{\frac12}\partial_z\overline{v}\|_{H^{k-1}}.$$

By Lemma 2.10, we have

$$\|T_{\partial_z\overline{v}}\overline{\gamma}\|_{H^{k-1}} \leq C\|\partial_z v\|_{L^\infty(\overline{\mathcal{S}})}\|\gamma\|_{H^{k-1}(\overline{\mathcal{S}})},$$

and for $k > 1$, we have

$$\|R(\partial_z\overline{v},\overline{\gamma})\|_{H^{k-1}} \leq C\|\partial_z v\|_{L^\infty(\overline{\mathcal{S}})}\|\gamma\|_{H^{k-1}(\overline{\mathcal{S}})}.$$

This proves that for $k > 1$,

$$\|\overline{\gamma}\partial_z\overline{v}\|_{H^{k-1}} \leq C\|\partial_z v\|_{L^\infty(\overline{\mathcal{S}})}\|\gamma\|_{H^{k-1}(\overline{\mathcal{S}})} + C\|\gamma\|_{L^\infty_z(I;C^{-\frac12})}\|\langle D\rangle^{\frac12}\partial_z v\|_{H^{k-1}(\overline{\mathcal{S}})}.$$

For $k = 1$, it is obvious that

$$\|\overline{\gamma}\partial_z\overline{v}\|_{L^2} \leq C\|\gamma\|_{L^2_z(I;L^\infty)}\|\partial_z v\|_{L^\infty_z(I;L^2)}.$$

The above estimates together with Lemma 4.4 and Lemma 4.5 ensure that

$$\begin{aligned}&\|\partial_z^{k-1}(\alpha\Delta v + \beta\cdot\nabla\partial_z v - \gamma\partial_z v)\|_{L^2(\overline{\mathcal{S}})}\\ &\leq C\|\nabla_{x,z}v\|_{L^\infty(\overline{\mathcal{S}})}\big(\|(\alpha-1,\beta)\|_{H^k(\overline{\mathcal{S}})} + \|\gamma\|_{H^{k-1}(\overline{\mathcal{S}})}\big)\\ &\quad + C\|(\alpha,\beta)\|_{L^\infty(\overline{\mathcal{S}})}\|\nabla\nabla_{x,z}v\|_{H^{k-1}(\overline{\mathcal{S}})}\\ &\quad + C\|\gamma\|_{L^\infty_z(I;C^{-\frac12})}\|\langle D\rangle^{\frac12}\partial_z v\|_{H^{k-1}(\overline{\mathcal{S}})} + C\|\gamma\|^2_{L^2_z(I;L^\infty)}\|\partial_z v\|_{L^\infty_z(I;L^2)}\\ &\leq C\|\nabla_{x,z}v\|_{L^\infty(\overline{\mathcal{S}})}\|\eta\|_{H^s} + K_\eta\big(\|\nabla\nabla_{x,z}v\|_{H^{k-1}(\overline{\mathcal{S}})}\\ &\quad + \|\langle D\rangle^{\frac12}\partial_z v\|_{H^{k-1}(\overline{\mathcal{S}})} + \|\partial_z v\|_{L^\infty_z(I;L^2)}\big),\end{aligned}$$

from which and (4.22), we infer that

$$\begin{aligned}\|\partial_z^{k+1}v\|_{L^2(\overline{\mathcal{S}})} &\leq C\|\nabla_{x,z}v\|_{L^\infty(\overline{\mathcal{S}})}\|\eta\|_{H^s}\\ &\quad + K_\eta\big(\|\nabla\nabla_{x,z}v\|_{H^{k-1}(\overline{\mathcal{S}})} + \|\langle D\rangle^{\frac12}\partial_z v\|_{H^{k-1}(\overline{\mathcal{S}})} + \|\partial_z v\|_{L^\infty_z(I;L^2)}\big)\\ &\quad + \|\partial_z^{k-1}F_0\|_{L^2(\overline{\mathcal{S}})}.\end{aligned}$$

which along with (4.20) and (4.21) implies the proposition by the interpolation. $\square$

4.5. Elliptic estimates in Besov space

In this subsection, we present the elliptic estimates in Besov space which will be used in the proof of break-down criterion.

PROPOSITION 4.16. *Let $I_0 = [a, 0]$ and $I_1 = [b, 0]$ with $b < a$. Let $q \in [1, \infty], r \in [0, \frac{1}{2}]$. Assume that $v \in X_q^r(I_1)$ is a solution of the elliptic equation*

$$\partial_z^2 v + \alpha \Delta v + \beta \cdot \nabla \partial_z v - \gamma \partial_z v = F_{00} + \partial_z G_0 \quad in\ \overline{\mathcal{S}}$$

with $v(x, 0) = f(x)$. Then it holds that for any $\delta > 0$,

$$\begin{aligned}\|\nabla_{x,z} v\|_{X_q^r(I_0)} \le K_\eta \Big(&\|f\|_{B^{r+1}_{\infty,q}} + \|F_{00}\|_{Y_q^r(I)} + \|G_0\|_{X_q^r(I)} \\ &+ \|\nabla_{x,z} v\|_{L_z^\infty(I_0;C^{-\delta})} + \|\nabla_{x,z} v\|_{\widetilde{L}_z^2(I_1 \setminus I_0;C^{-\frac{\epsilon}{2}})}\Big).\end{aligned}$$

Here $X_q^r(I_0) = \widetilde{L}_z^\infty(I_0; B^r_{\infty,q})$.

Let us recall the following lemma from **[33]**, which can be proved by using Bony's decomposition (2.3), Lemma 2.17 and Lemma 2.18.

LEMMA 4.17. *Let $I_0 = [a, 0]$ and $I_1 = [b, 0]$ with $b < a$. Then for any $r \le \frac{1}{2}$ and $q \in [1, \infty]$, it holds that*

$$\begin{aligned}\|F_1\|_{\widetilde{L}_z^2(I_1;B^{r-\frac{1}{2}}_{\infty,q})} \le &K_\eta \big(\|\nabla_{x,z} v\|_{\widetilde{L}_z^\infty(I_0;C^{-\frac{\epsilon}{2}})} + \|\partial_z v\|_{\widetilde{L}_z^\infty(I_0;B^{r-\frac{1}{2}}_{\infty,q})} \\ &+ \|\nabla_{x,z} v\|_{\widetilde{L}_z^2(I_1 \setminus I_0;C^{-\frac{\epsilon}{2}})}\big),\\ \|F_2\|_{\widetilde{L}_z^2(I_1;B^{r-\frac{1}{2}}_{\infty,q})} \le &K_\eta \big(\|\nabla_{x,z} v\|_{\widetilde{L}_z^\infty(I_0;C^{-\frac{\epsilon}{2}})} + \|\nabla_{x,z} v\|_{\widetilde{L}_z^2(I_1 \setminus I_0;C^{-\frac{\epsilon}{2}})}\big),\\ \|F_3\|_{\widetilde{L}_z^2(I_1;B^{r-\frac{1}{2}}_{\infty,q})} \le &K_\eta \big(\|\nabla_{x,z} v\|_{\widetilde{L}_z^\infty(I_0;C^{-\frac{\epsilon}{2}})} + \|\nabla_{x,z} v\|_{\widetilde{L}_z^2(I_1 \setminus I_0;C^{-\frac{\epsilon}{2}})}\big).\end{aligned}$$

PROOF OF PROPOSITION 4.16. Let χ be a smooth function supported in I_1 and $\chi(z) = 1$ for $z \in I_0$. Set $w = \chi(z)(\partial_z - T_A)v$. Then (w, v) satisfies

$$\begin{aligned}&\partial_z w - T_a w = \widetilde{F}, \quad w(b) = 0,\\ &\partial_z v - T_A v = w \quad \text{in } [a, 0], \quad v(0) = f,\end{aligned}$$

where $\widetilde{F} = \chi(z)\big(F_{00} + \partial_z G_0 + F_1 + F_2 + F_3\big) + \chi'(z)(\partial_z - T_A)v$.

Let $w_1 = w - \chi(z) G_0$. Then w_1 satisfies

$$\partial_z w_1 - T_a w_1 = \widetilde{F} - \partial_z(\chi(z) G_0) - \chi(z) T_a G_0, \quad w(b) = 0.$$

Then Proposition 3.2 together with Lemma 4.17 and Proposition 2.7 ensures that

$$\begin{aligned}\|w - G_0\|_{X^r_{2,q}(I_0)} \le &K_\eta \big(\|\widetilde{F} - \chi(z)\partial_z G_0\|_{Y_q^r(I_1)} + \|G_0\|_{X_q^r(I_1)} + \|w\|_{\widetilde{L}_z^2(I_1;C^{-\delta})}\big)\\ \le &K_\eta \big(\|F_{00}\|_{Y_q^r(I)} + \|G_0\|_{X_q^r(I_1)} + \|\nabla_{x,z} v\|_{\widetilde{L}_z^\infty(I_0;B^{r-\frac{1}{2}}_{\infty,q})} \\ &+ \|\nabla_{x,z} v\|_{\widetilde{L}_z^\infty(I_0;C^{-\frac{\epsilon}{2}})} + \|\nabla_{x,z} v\|_{\widetilde{L}_z^2(I_0 \setminus I_0;C^{-\frac{\epsilon}{2}})}\big).\end{aligned}$$

Note that $(\partial_z - T_A)\nabla v = \nabla(w-G_0)+\nabla G_0+T_{\nabla A}v$ on I_0. It follows from Proposition 3.2, Lemma 4.6 and Remark 2.8 that

$$\begin{aligned}\|\nabla v\|_{X^r_q(I_0)} \le K_\eta\big(&\|f\|_{B^{r+\frac12}_{\infty,q}} + \|w-G_0\|_{\widetilde{L}^2_z(I_0;B^{r+\frac12}_{\infty,q})} + \|G_0\|_{X^r_q(I_1)}\\ &+ \|\nabla_{x,z}v\|_{\widetilde{L}^\infty_z(I_0;B^{r-\frac12}_{\infty,q})} + \|\nabla_{x,z}v\|_{\widetilde{L}^\infty_z(I_0;C^{-\delta})}\big)\\ \le K_\eta\big(&\|f\|_{B^{r+\frac12}_{\infty,q}} + \|F_{00}\|_{Y^r_q(I)} + \|G_0\|_{X^r_q(I_1)} + \|\nabla_{x,z}v\|_{\widetilde{L}^\infty_z(I_0;B^{r-\frac12}_{\infty,q})}\\ &+ \|\nabla_{x,z}v\|_{\widetilde{L}^\infty_z(I_0;C^{-\frac{\varepsilon}{2}})} + \|\nabla_{x,z}v\|_{\widetilde{L}^2_z(I_1\setminus I_0;C^{-\frac{\varepsilon}{2}})}\big).\end{aligned}$$

The same estimate for $\partial_z v$ can be deduced by using $\partial_z v = T_A v + w$ and Proposition 2.7. Thus, we obtain

$$\begin{aligned}\|\nabla_{x,z}v\|_{X^r_q(I_0)} \le &K_\eta\big(\|f\|_{B^{r+\frac12}_{\infty,q}} + \|F_{00}\|_{Y^r_q(I)} + \|G_0\|_{X^r_q(I_1)} + \|\nabla_{x,z}v\|_{\widetilde{L}^\infty_z(I_0;B^{r-\frac12}_{\infty,q})}\\ &+ \|\nabla_{x,z}v\|_{\widetilde{L}^\infty_z(I_0;C^{-\frac{\varepsilon}{2}})} + \|\nabla_{x,z}v\|_{\widetilde{L}^2_z(I_1\setminus I_0;C^{-\frac{\varepsilon}{2}})}\big).\end{aligned}$$

This together with the interpolation inequality

$$\|\nabla_{x,z}v\|_{\widetilde{L}^\infty_z(I_0;B^{r-\frac12}_{\infty,q}\cap C^{-\frac{\varepsilon}{2}})} \le K_\eta\|\nabla_{x,z}v\|_{L^\infty_z(I_0;C^{-\delta})} + \frac{1}{4K_\eta}\|\nabla_{x,z}v\|_{\widetilde{L}^\infty_z(I_0;B^r_{\infty,q})}$$

implies the proposition. □

4.6. Interior $W^{1,p}$ estimate

We consider the elliptic equation

$$\Delta_{x,y}\phi = \operatorname{div}_{x,y} g \quad \text{in} \quad \mathcal{S}. \tag{4.23}$$

Given a point $X_0 = (x_0, y_0) \in \mathbf{R}^d \times \mathbf{R}$, let $B_r(X_0)$ be a ball of radius r centered at X_0. We have the following interior $W^{1,p}$ estimate.

PROPOSITION 4.18. *Suppose that $\phi \in H^1(\mathcal{S})$ is a solution of (4.23) with $g \in L^p(\mathcal{S})$ for $p \ge 2$. Let $I_x = [-1+c_1h_0, \eta(x)-c_2h_0]$, where $c_1, c_2 > 0$ and $c_1+c_2 < 1$. Then there exits $\delta_1 > 0$ depending only on $\|\eta\|_{C^{1+\varepsilon}}$, h_0 and c_1, c_2 so that $B_{\delta_1}(X_0) \subseteq \mathcal{S}$ for any $X_0 \in \{(x,y) : x \in \mathbf{R}^d, y \in I_x\}$ and*

$$\|\phi\|_{W^{1,p}(B_{\delta_1/2}(X_0))} \le K_\eta\big(\|g\|_{L^p(B_{\delta_1}(X_0))} + \|\nabla_{x,y}\phi\|_{L^2(B_{\delta_1}(X_0))}\big).$$

PROOF. Given any point $X_0 \in \mathbf{R}^d \times I_x$. Let $\overline{\phi}$ be a solution of the elliptic equation in $B_{\delta_1}(X_0)$:

$$\Delta_{x,y}\overline{\phi} = \operatorname{div}_{x,y} g \quad \text{in } B_{\delta_1}(X_0), \quad \overline{\phi}|_{\partial B_{\delta_1}(X_0)} = 0.$$

The classical $W^{1,p}$ elliptic estimate ensures that

$$\|\overline{\phi}\|_{W^{1,p}(B_{\delta_1}(X_0))} \le C(\delta_1)\|g\|_{L^p(B_{\delta_1}(X_0))}.$$

Then by using the interior gradient estimate of the harmonic function $\phi - \overline{\phi}$, we deduce that

$$\begin{aligned}\|\phi\|_{W^{1,p}(B_{\delta_1/2}(X_0))} \le &\|\phi-\overline{\phi}\|_{W^{1,p}(B_{\delta_1/2}(X_0))} + \|\overline{\phi}\|_{W^{1,p}(B_{\delta_1/2}(X_0))}\\ \le &C(\delta_1)\big(\|\nabla_{x,y}(\phi-\overline{\phi})\|_{L^2(B_{\delta_1}(X_0))} + \|\overline{\phi}\|_{W^{1,p}(B_{\delta_1}(X_0))}\big)\\ \le &C(\delta_1)\big(\|\nabla_{x,y}\phi\|_{L^2(B_{\delta_1}(X_0))} + \|g\|_{L^p(B_{\delta_1}(X_0))}\big).\end{aligned}$$

The proof is finished. □

CHAPTER 5

Dirichlet-Neumann operator

In this chapter, we assume that $\eta \in C^{\frac{3}{2}+\varepsilon}(\mathbf{R}^d)$ for some $\varepsilon > 0$ and satisfies (4.1). We will use some notations from chapter 4.

5.1. Definition and paralinearization

We consider the boundary value problem

$$\left\{ \begin{array}{ll} \Delta_{x,y}\phi = 0 & \text{in} \quad \mathcal{S}, \\ \phi|_{y=\eta(x)} = f, & \phi|_{y=-1} = 0. \end{array} \right. \tag{5.1}$$

where $\mathcal{S} = \big\{(x, y) : x \in \mathbf{R}^d, -1 < y < \eta(x)\big\}$. Given $f \in H^{\frac{1}{2}}(\mathbf{R}^d)$, the existence of the variation solution ϕ with $\nabla_{x,y}\phi \in L^2(\mathcal{S})$ can be deduced by using Riesz theorem, see [**3**] for example. Moreover, it holds that

$$\|\nabla_{x,y}\phi\|_{L^2(\mathcal{S})} \le C(\|\eta\|_{W^{1,\infty}}, h_0)\|f\|_{H^{\frac{1}{2}}}. \tag{5.2}$$

Let $\eta \in H^s(\mathbf{R}^d)$ for $s > \frac{d}{2} + \frac{3}{2}$. This together with Proposition 4.10 yields that for any $\sigma \in [-\frac{1}{2}, s-1]$,

$$\|\widetilde{\phi}\|_{H^{\sigma+\frac{3}{2}}(\overline{\mathcal{S}})} \le P_{s,\eta}\|f\|_{H^{\sigma+1}}. \tag{5.3}$$

DEFINITION 5.1. Given η, f, ϕ as above, the Dirichlet-Neumann(DN) operator $G(\eta)$ is defined by

$$G(\eta)f \stackrel{\text{def}}{=} \sqrt{1+|\nabla\eta|^2}\partial_n\phi\big|_{y=\eta}.$$

The DN operator is a positive self-adjoint operator. More precisely, for any $f, g \in H^{\frac{1}{2}}(\mathbf{R}^d)$, we have

$$\begin{aligned} &\langle G(\eta)f, g\rangle = \langle f, G(\eta)g\rangle, \\ &\langle G(\eta)f, f\rangle = \|\nabla_{x,y}\phi\|^2_{L^2(\mathcal{S})} \ge c\|f\|^2_{H^{\frac{1}{2}}}. \end{aligned}$$

In terms of $\widetilde{\phi}$, the Dirichlet-Neumann operator $G(\eta)$ can be written as

$$G(\eta)f = \Big(\frac{1+|\nabla\rho_\delta|^2}{\partial_z\rho_\delta}\partial_z\widetilde{\phi} - \nabla\rho_\delta\cdot\nabla\widetilde{\phi}\Big)\Big|_{z=0}.$$

Following [**3**], we first paralinearize $G(\eta)$. We set

$$\zeta_1 = \frac{1+|\nabla\rho_\delta|^2}{\partial_z\rho_\delta}\Big|_{z=0}, \quad \zeta_2 = \nabla\rho_\delta\big|_{z=0}.$$

It is easy to show that

$$\|\zeta_1 - 1\|_{H^{s-1}} + \|\zeta_2\|_{H^{s-1}} \le K_\eta\|\eta\|_{H^s}. \tag{5.4}$$

Using Bony's decomposition (2.3), we decompose $G(\eta)$ as

$$\begin{aligned}G(\eta)f =&\partial_z\widetilde{\phi} + T_{\zeta_1-1}\partial_z\widetilde{\phi} + T_{\partial_z\widetilde{\phi}}(\zeta_1-1) + R(\zeta_1-1,\partial_z\widetilde{\phi}) - T_{i\zeta_2\cdot\xi}\widetilde{\phi}\\ &- T_{\nabla\widetilde{\phi}}\cdot\zeta_2 - R(\zeta_2,\nabla\widetilde{\phi})\big|_{z=0}.\end{aligned}$$

Replacing $\partial_z\widetilde{\phi}$ by $T_A\widetilde{\phi}$, we get

$$G(\eta)f = T_\lambda f + R(\eta)f, \tag{5.5}$$

where $\lambda = \zeta_1 A - i\zeta_2\cdot\xi\big|_{z=0}$ with

$$A = \frac{1}{2}(-i\beta\cdot\xi + \sqrt{4\alpha|\xi|^2 - (\beta\cdot\xi)^2}), \tag{5.6}$$

and $R(\eta)$ is the remainder of DN operator given by

$$\begin{aligned}R(\eta)f =&\Big[(T_{\zeta_1}T_A - T_{\zeta_1 A})\widetilde{\phi} - T_{\zeta_1}(\partial_z - T_A)\widetilde{\phi}\\ &+ \big(S_2(\partial_z\widetilde{\phi}) + T_{\partial_z\widetilde{\phi}}(\zeta_1-1) + R(\zeta_1-1,\partial_z\widetilde{\phi}) - T_{\nabla\widetilde{\phi}}\cdot\zeta_2 - R(\nabla\widetilde{\phi},\zeta_2)\big)\Big]\Big|_{z=0}\\ \triangleq& R_1(\eta)f + R_2(\eta)f + R_3(\eta)f.\end{aligned} \tag{5.7}$$

where

$$\begin{aligned}R_1(\eta)f &= (T_{\zeta_1}T_A - T_{\zeta_1 A})\widetilde{\phi}\Big|_{z=0},\\ R_2(\eta)f &= -T_{\zeta_1}(\partial_z - T_A)\widetilde{\phi}\Big|_{z=0},\\ R_3(\eta)f &= \big(S_2(\partial_z\widetilde{\phi}) + T_{\partial_z\widetilde{\phi}}(\zeta_1-1) + R(\zeta_1-1,\partial_z\widetilde{\phi}) - T_{\nabla\widetilde{\phi}}\cdot\zeta_2 - R(\nabla\widetilde{\phi},\zeta_2)\big)\Big|_{z=0}.\end{aligned}$$

Such paralinearization of the Dirichlet-Neumann operator was first introduced in [**4**].

5.2. Sobolev estimate of the remainder

In this section, we present Sobolev estimates for the remainder in the case when the boundary is smooth, which will used in the proof of the uniform estimates of the approximate solutions.

PROPOSITION 5.2. *Assume that $\eta\in H^s(\mathbf{R}^d)$ for $s>\frac{d}{2}+2$. Then it holds that*

$$\|R(\eta)f\|_{H^{s-1}} \le P_{s,\eta}\|f\|_{H^{s-1}}.$$

PROOF. Thanks to the fact that for $s>\frac{d}{2}+2$,

$$M_1^0(\zeta_1) + \mathcal{M}_1^1(A) \le P_{s,\eta},$$

we deduce from Proposition 2.6 and (5.3) that

$$\|R_1(\eta)f\|_{H^{s-1}} \le P_{s,\eta}\|\nabla\widetilde{\phi}\|_{L_z^\infty(I;H^{s-2})} \le P_{s,\eta}\|f\|_{H^{s-1}}.$$

By Lemma 2.10 and (5.4), we have

$$\begin{aligned}\|R_3(\eta)f\|_{H^{s-1}} \le& C\|\nabla_{x,z}\widetilde{\phi}\|_{L_z^\infty(I;L^2)} + P_{s,\eta}\|\nabla_{x,z}\widetilde{\phi}\|_{L^\infty(\overline{\mathcal{S}})}\\ \le& P_{s,\eta}\|\nabla_{x,z}\widetilde{\phi}\|_{L_z^\infty(I;H^{s-2})} \le P_{s,\eta}\|f\|_{H^{s-1}}.\end{aligned}$$

For $R_2(\eta)$, we get by Lemma 2.10 that

$$\|R_2(\eta)f\|_{H^{s-1}} \leq P_{s,\eta}\|(\partial_z - T_A)\widetilde{\phi}\|_{L_z^\infty([-\frac{1}{2},0];H^{s-1})}. \tag{5.8}$$

While by the proof of Proposition 4.12, we know that $w = \chi(z)(\partial_z - T_A)\widetilde{\phi}$ satisfies

$$\partial_z w - T_a w = F', \quad w(-1) = 0.$$

where $F' = \chi(z)\big(F_1 + F_2 + F_3\big) + \chi'(z)(\partial_z - T_A)\widetilde{\phi}$ and χ is a smooth function satisfying $\chi(-1) = 0$ and $\chi(z) = 1$ for $z \in [-\frac{1}{2}, 0]$. Then it follows from Proposition 3.1, Lemma 5.3 and (5.3) that

$$\begin{aligned}
\|(\partial_z - T_A)\widetilde{\phi}\|_{X^{s-1}([-\frac{1}{2},0])} \leq& \|w\|_{X^{s-1}(I)} \\
\leq& P_{s,\eta}\Big(\sum_{i=1}^{3}\|F_i\|_{L_z^2(I;H^{s-\frac{3}{2}})} + \|\nabla_{x,z}\widetilde{\phi}\|_{X^{s-2}(I)}\Big) \\
\leq& P_{s,\eta}\|f\|_{H^{s-1}},
\end{aligned}$$

which along with (5.8) gives

$$\|R_2(\eta)f\|_{H^{s-1}} \leq P_{s,\eta}\|f\|_{H^{s-1}}.$$

Putting the estimates of $R_i(\eta)f$ together concludes the proposition. □

LEMMA 5.3. *Let $s > \frac{d}{2} + 2$. It holds that for $i = 1, 2, 3$,*

$$\|F_i\|_{L_z^2(I;H^{s-\frac{3}{2}})} \leq P_{s,\eta}\|\nabla_{x,z}\widetilde{\phi}\|_{X^{s-2}(I)}.$$

PROOF. Recall that $F_1 = \gamma\partial_z\widetilde{v}$. It follows from Lemma 2.13 that

$$\begin{aligned}
\|F_1\|_{L_z^2(I;H^{s-\frac{3}{2}})} \leq& C\|\gamma\|_{L^\infty}\|\partial_z\widetilde{\phi}\|_{L_z^2(I;H^{s-\frac{3}{2}})} + C\|\partial_z\widetilde{\phi}\|_{L^\infty}\|\gamma\|_{L_z^2(I;H^{s-\frac{3}{2}})} \\
\leq& C\|\gamma\|_{L_z^\infty(I;H^{s-2})}\|\partial_z\widetilde{\phi}\|_{L_z^2(I;H^{s-\frac{3}{2}})} + C\|\partial_z\widetilde{\phi}\|_{L_z^\infty(I;H^{s-2})}\|\gamma\|_{L_z^2(I;H^{s-\frac{3}{2}})} \\
\leq& P_{s,\eta}\|\nabla_{x,z}\widetilde{\phi}\|_{X^{s-2}(I)}.
\end{aligned}$$

Recall that

$$F_2 = (T_\alpha - \alpha)\Delta\widetilde{\phi} + (T_\beta - \beta)\cdot\nabla\partial_z\widetilde{\phi}.$$

Then we get by Lemma 2.10 that

$$\begin{aligned}
\|F_2\|_{L_z^2(I;H^{s-\frac{3}{2}})} \leq& C\|\nabla_{x,z}\widetilde{\phi}\|_{L^\infty}\big(1 + \|\alpha - 1\|_{L_z^2(I;H^{s-\frac{1}{2}})} + \|\beta\|_{L_z^2(I;H^{s-\frac{1}{2}})}\big) \\
\leq& P_{s,\eta}\|\nabla_{x,z}\widetilde{\phi}\|_{X^{s-2}(I)}.
\end{aligned}$$

Recall that

$$F_3 = (T_a T_A - T_\alpha\Delta)\widetilde{\phi} - (T_a + T_A + T_\beta\cdot\nabla)\partial_z\widetilde{\phi} - T_{\partial_z A}\widetilde{\phi}.$$

For $s > \frac{d}{2} + 2$, we have

$$\mathcal{M}_1^0(a) + \mathcal{M}_1^1(A) \leq P_{s,\eta}.$$

Then we deduce from Lemma 4.6 and Proposition 2.6 that

$$\|F_3\|_{L_z^2(I;H^{s-\frac{3}{2}})} \leq P_{s,\eta}\|\nabla_{x,z}\widetilde{\phi}\|_{X^{s-2}(I)}.$$

The proof is finished. □

5.3. Tame estimate of the remainder

In this section, we present tame estimates for the remainder in the case when the boundary has more one half derivative than f. The result will be used in the proof of the break-down criterion.

PROPOSITION 5.4. *Assume that $\eta \in H^{s+\frac{1}{2}}(\mathbf{R}^d)$ for $s > \frac{d}{2}+1$. Let I_1 be as in Proposition 4.12 and $\widetilde{S} = \mathbf{R}^d \times I_1$. It holds that*

$$\|R(\eta)f\|_{H^{s-\frac{1}{2}}} \leq K_\eta\big(\|f\|_{H^s} + \|\nabla_{x,z}\widetilde{\phi}\|_{L^\infty(\widetilde{S})}\|\eta\|_{H^{s+\frac{1}{2}}}\big),$$
$$\|R(\eta)f\|_{H^{s-1}} \leq K_\eta\big(\|f\|_{H^{s-\frac{1}{2}}} + \|\nabla_{x,z}\widetilde{\phi}\|_{L^\infty_z(I_1;C^0)}\|\eta\|_{H^{s+\frac{1}{2}}}\big).$$

PROOF. Note that $A \in \Gamma^1_{\frac{1}{2}+\varepsilon}(I \times \mathbf{R}^d)$ and $\zeta_1 \in \Gamma^0_{\frac{1}{2}+\varepsilon}(\mathbf{R}^d)$ with the bound

$$\mathcal{M}^1_{\frac{1}{2}+\varepsilon}(A) + M^0_{\frac{1}{2}+\varepsilon}(\zeta_1) \leq K_\eta. \tag{5.9}$$

Then we get by Proposition 2.6, (5.4), Proposition 4.12 and (5.2) that

$$\begin{aligned}\|R_1(\eta)f\|_{H^{s-\frac{1}{2}}} \leq& K_\eta\|\nabla\widetilde{\phi}\|_{L^\infty_z([z_0,0];H^{s-1})}\\ \leq& K_\eta\big(\|\nabla_{x,z}\widetilde{\phi}\|_{L^2(\widetilde{S})} + \|f\|_{H^s} + \|\nabla_{x,z}\widetilde{\phi}\|_{L^\infty_z(I_1;C^0)}\|\eta\|_{H^{s+\frac{1}{2}}}\big)\\ \leq& K_\eta\big(\|f\|_{H^s} + \|\nabla_{x,z}\widetilde{\phi}\|_{L^\infty_z(I_1;C^0)}\|\eta\|_{H^{s+\frac{1}{2}}}\big).\end{aligned}$$

Similarly, we deduce from Lemma 2.10 that

$$\begin{aligned}\|R_3(\eta)f\|_{H^{s-\frac{1}{2}}} \leq& K_\eta\|\nabla_{x,z}\widetilde{\phi}\|_{L^\infty(\widetilde{S})}\|\eta\|_{H^{s+\frac{1}{2}}} + \|\partial_z\widetilde{\phi}\|_{L^\infty_z([z_0,0];H^{s-1})}\\ \leq& K_\eta\big(\|f\|_{H^s} + \|\nabla_{x,z}\widetilde{\phi}\|_{L^\infty(\widetilde{S})}\|\eta\|_{H^{s+\frac{1}{2}}}\big).\end{aligned}$$

While by the proof of (4.19), we see that

$$\begin{aligned}\|(\partial_z - T_A)\widetilde{\phi}\|_{L^\infty_z([z_0,0];H^{s-\frac{1}{2}})} \leq& K_\eta\big(\|\nabla_{x,z}\widetilde{\phi}\|_{X^{s-1}([z_0,0])} + \|\nabla_{x,z}\widetilde{\phi}\|_{L^\infty(\widetilde{S})}\|\eta\|_{H^{s+\frac{1}{2}}}\big)\\ \leq& K_\eta\big(\|f\|_{H^s} + \|\nabla_{x,z}\widetilde{\phi}\|_{L^\infty(\widetilde{S})}\|\eta\|_{H^{s+\frac{1}{2}}}\big).\end{aligned}$$

Then we get by Lemma 2.10 that

$$\begin{aligned}\|R_2(\eta)f\|_{H^{s-\frac{1}{2}}} \leq& K_\eta\|(\partial_z - T_A)\widetilde{\phi}\|_{L^\infty_z([z_0,0];H^{s-\frac{1}{2}})}\\ \leq& K_\eta\big(\|f\|_{H^s} + \|\nabla_{x,z}\widetilde{\phi}\|_{L^\infty(\widetilde{S})}\|\eta\|_{H^{s+\frac{1}{2}}}\big).\end{aligned}$$

This completes the proof of the first inequality. The second inequality can be proved similarly. □

5.4. Hölder estimate of the remainder

PROPOSITION 5.5. *Let $I_0 = [-\frac{1}{2}, 0]$ and $I_1 = [-\frac{3}{4}, 0]$. It holds that for any $\delta > 0$,*

$$\|R(\eta)f\|_{C^{\frac{1}{2}}} \leq K_\eta\big(\|f\|_{C^1} + \|\nabla_{x,z}\widetilde{\phi}\|_{\widetilde{L}^\infty_z(I_0;C^{-\delta})} + \|\nabla_{x,z}\widetilde{\phi}\|_{\widetilde{L}^2_z(I_1\setminus I_0;C^{-\frac{\varepsilon}{2}})}\big).$$

PROOF. Thanks to (5.9), we get by Proposition 2.7 that

$$\|R_1(\eta)f\|_{C^{\frac{1}{2}}} \leq K_\eta\|\nabla\widetilde{\phi}\|_{L^\infty_z(I_0;C^0)}.$$

We can see from the proof of Proposition 4.16 that

$$\|(\partial_z - T_A)\widetilde{\phi}\|_{\widetilde{L}^\infty_z([-\frac{1}{4},0];C^{\frac{1}{2}})} \leq K_\eta(\|\nabla_{x,z}\widetilde{\phi}\|_{\widetilde{L}^\infty_z(I_0;C^0)} + \|\nabla_{x,z}\widetilde{\phi}\|_{\widetilde{L}^2_z(I_1\setminus I_0;C^{-\frac{\varepsilon}{2}})}),$$

which together with Lemma 2.10 gives

$$\begin{aligned}\|R_2(\eta)f\|_{C^{\frac{1}{2}}} &\leq K_\eta\|(\partial_z - T_A)\widetilde{\phi}\|_{L^\infty_z([-\frac{1}{4},0];C^{\frac{1}{2}})}\\ &\leq K_\eta(\|\nabla_{x,z}\widetilde{\phi}\|_{\widetilde{L}^\infty_z(I_0;C^0)} + \|\nabla_{x,z}\widetilde{\phi}\|_{\widetilde{L}^2_z(I_1\setminus I_0;C^{-\frac{\epsilon}{2}})}).\end{aligned}$$

By Lemma 2.10 again, we get

$$\|R_3(\eta)f\|_{C^{\frac{1}{2}}} \leq K_\eta\|\nabla_{x,z}\widetilde{\phi}\|_{L^\infty_z(I;C^0)}.$$

This together with Proposition 4.16 shows that

$$\begin{aligned}\|R(\eta)f\|_{C^{\frac{1}{2}}} &\leq K_\eta(\|\nabla_{x,z}\widetilde{\phi}\|_{\widetilde{L}^\infty_z(I_0;C^0)} + \|\nabla_{x,z}\widetilde{\phi}\|_{\widetilde{L}^2_z(I_1\setminus I_0;C^{-\frac{\epsilon}{2}})})\\ &\leq K_\eta\big(\|f\|_{B^1_{\infty,\infty}} + \|\nabla_{x,z}\widetilde{\phi}\|_{\widetilde{L}^\infty_z(I_0;C^{-\delta})} + \|\nabla_{x,z}\widetilde{\phi}\|_{\widetilde{L}^2_z(I_1\setminus I_0;C^{-\frac{\epsilon}{2}})}\big).\end{aligned}$$

This finishes the proof. □

CHAPTER 6

New formulation and paralinearization

6.1. New formulation

First of all, we derive the evolution equations for the free surface and the trace of the velocity on the boundary. We denote

$$V \triangleq (v^1, \cdots, v^d)\big|_{y=\eta}, \quad V_b \triangleq (v^1, \cdots, v^d)\big|_{y=-1}, \quad B \triangleq v^{d+1}\big|_{y=\eta},$$
$$a \triangleq -\partial_y P|_{y=\eta}, \quad \zeta \triangleq \nabla\eta.$$

Using the fact that for any function $f = f(t, x, y)$,

$$(\partial_t + V\cdot\nabla)(f|_{y=\eta}) = \big(\partial_t f + v\cdot\nabla_{x,y} f\big)\big|_{y=\eta},$$

we deduce from (1.1) that

$$(\partial_t + V\cdot\nabla)B = a - 1, \tag{6.1}$$
$$(\partial_t + V\cdot\nabla)V + a\zeta = 0, \tag{6.2}$$
$$(\partial_t + V_b\cdot\nabla)V_b + \nabla P|_{y=-1} = 0. \tag{6.3}$$

Let ω be the vorticity of the fluid, which is defined by

$$\omega = \big(\omega_{i,j}\big)_{1\le i,j\le d+1}, \quad \omega_{i,j} = \partial_{x_i} v^j - \partial_{x_j} v^i.$$

The motion of the fluid is determined by the vorticity equation

$$\omega_t + v\cdot\nabla_{x,y}\omega = \omega\otimes\nabla_{x,y} v \quad \text{in} \quad \Omega_t. \tag{6.4}$$

Here $(\omega\otimes\nabla_{x,y}v)_{i,j} = \omega_{k,i}\partial_j v^k + \omega_{k,j}\partial_i v^k$.

The velocity v can be recovered from the vorticity ω by solving the elliptic equation

$$\begin{cases} \Delta_{x,y} v = -\nabla_{x,y}\times\omega \quad \text{in} \quad \Omega_t, \\ v|_{y=\eta} = (V, B), \quad v|_{y=-1} = (V_b, 0). \end{cases} \tag{6.5}$$

where $\omega = \big(\omega_{i,j}\big)_{1\le i,j\le d+1}$ with $\omega_{i,j} = \partial_{x_i} v^j - \partial_{x_j} v^i$. The pressure P of the fluid is determined by solving the following elliptic equation:

$$\begin{cases} -\Delta_{x,y} P = \partial_i v^j \partial_j v^i \quad \text{in} \quad \Omega_t, \\ P|_{y=\eta} = 0, \quad \partial_y P|_{y=-1} = -1. \end{cases} \tag{6.6}$$

We decompose the velocity v into the irrotational part v_{ir} and the rotational part v_ω, i.e.,

$$\begin{cases} \Delta_{x,y} v_{ir} = 0 \quad \text{in} \quad \Omega_t, \\ v_{ir}|_{y=\eta} = (V, B), \quad v_{ir}|_{y=-1} = 0, \end{cases} \tag{6.7}$$

and

$$\begin{cases} \Delta_{x,y} v_\omega = -\nabla_{x,y}\times\omega \quad \text{in} \quad \Omega_t, \\ v_\omega|_{y=\eta} = 0, \quad v_\omega|_{y=-1} = (V_b, 0). \end{cases} \tag{6.8}$$

It follows from (1.4) that

$$(\partial_t + V\cdot\nabla)\zeta = \nabla B - \sum_j \nabla V_j \zeta^j. \tag{6.9}$$

Then a direct calculation yields

$$\begin{aligned}
&\partial_{x_i} B - \partial_{x_i} V_j \partial_{x_j}\eta \\
&= \partial_{x_i} v^{d+1} + \partial_{x_i}\eta\partial_y v^{d+1} - \partial_{x_j}\eta\big(\partial_{x_i} v^j + \partial_{x_i}\eta\partial_y v^j\big)\big|_{y=\eta} \\
&= \big(\partial_y v^i - \partial_{x_j} v^i\cdot\partial_{x_j}\eta\big) + \partial_{x_i}\eta\big(\partial_y v^{d+1} - \partial_{x_j}\eta\partial_{x_j} v^{d+1}\big) \\
&\quad + \big(\omega_{i,d+1} - \partial_{x_j}\eta\omega_{ij} + \partial_{x_i}\eta\partial_{x_j}\eta\omega_{j,d+1}\big)\big|_{y=\eta} \\
&= \big(\partial_y v^i_{ir} - \partial_{x_j} v^i_{ir}\cdot\partial_{x_j}\eta\big) + \partial_{x_i}\eta\big(\partial_y v^{d+1}_{ir} - \partial_{x_j}\eta\partial_{x_j} v^{d+1}_{ir}\big) \\
&\quad + \big(\partial_y v^i_{\omega} - \partial_{x_j} v^i_{\omega}\cdot\partial_{x_j}\eta\big) + \partial_{x_i}\eta\big(\partial_y v^{d+1}_{\omega} - \partial_{x_j}\eta\partial_{x_j} v^{d+1}_{\omega}\big) \\
&\quad + \big(\omega_{i,d+1} - \partial_{x_j}\eta\omega_{ij} + \partial_{x_i}\eta\partial_{x_j}\eta\omega_{j,d+1}\big)\big|_{y=\eta} \\
&= G(\eta)V_i + \partial_{x_i}\eta G(\eta)B + R^i_{\omega},
\end{aligned}$$

where

$$\begin{aligned}
R^i_{\omega} \triangleq &\big(\partial_y v^i_{\omega} - \partial_{x_j} v^i_{\omega}\cdot\partial_{x_j}\eta\big) + \partial_{x_i}\eta\big(\partial_y v^{d+1}_{\omega} - \partial_{x_j}\eta\partial_{x_j} v^{d+1}_{\omega}\big) \\
&+ \big(\omega_{i,d+1} - \partial_{x_j}\eta\omega_{ij} + \partial_{x_i}\eta\partial_{x_j}\eta\omega_{j,d+1}\big)\big|_{y=\eta}.
\end{aligned}$$

Thus, $\zeta = \nabla\eta$ satisfies

$$(\partial_t + V\cdot\nabla)\zeta = G(\eta)V + \zeta G(\eta)B + R_{\omega}. \tag{6.10}$$

The term R_ω induced by the vorticity will lead the system to lose one half derivative.

6.2. Paralinearization

We paralinearize the system (6.1), (6.2) and (6.10). For this end, we introduce so called good unknown $U = V + T_\zeta B$. Applying Bony's decomposition and (5.5) to (6.1), (6.2) and (6.10), we obtain

$$\left\{\begin{array}{l} (\partial_t + T_V\cdot\nabla)V + T_a\zeta + T_\zeta(\partial_t + T_V\cdot\nabla)B = h_1, \\ (\partial_t + T_V\cdot\nabla)\zeta = T_\lambda U + h_2 + R_\omega, \end{array}\right. \tag{6.11}$$

where

$$\begin{aligned}
h_1 &\triangleq (T_V - V)\cdot\nabla V - R(a,\zeta) + T_\zeta(T_V - V)\cdot\nabla B, \\
h_2 &\triangleq (T_V - V)\cdot\nabla\zeta + [T_\zeta, T_\lambda]B + (\zeta - T_\zeta)T_\lambda B + R(\eta)V + \zeta R(\eta)B.
\end{aligned}$$

Let $D_t \triangleq \partial_t + T_V\cdot\nabla$. In terms of good unknown, the first equation of (6.11) can be rewritten as

$$D_t U + T_a\zeta = h_1 + [D_t, T_\zeta]B. \tag{6.12}$$

Taking D_t on both sides of (6.12), we get

$$D_t^2 U + T_a D_t\zeta = D_t h_1 + [T_a, D_t]\zeta + D_t[D_t, T_\zeta]B,$$

which along with the second equation of (6.11) gives

$$D_t^2 U + T_{a\lambda} U = f + f_\omega, \tag{6.13}$$

where (f, f_ω) is given by

$$f \triangleq D_t h_1 + (T_{a\lambda} - T_\lambda T_a)U + [T_a, D_t]\zeta + D_t[D_t, T_\zeta]B - T_a h_2, \quad f_\omega \triangleq -T_a R_\omega.$$

Similarly, we have

$$D_t^2\zeta + T_{a\lambda}\zeta = g + g_\omega, \tag{6.14}$$

where (g, g_ω) is given by

$$g \triangleq D_t h_2 + [D_t, T_\lambda]U + T_\lambda(h_1 + D_t[D_t, T_\zeta]B) + (T_{a\lambda} - T_\lambda T_a)\zeta, \quad g_\omega \triangleq D_t R_\omega.$$

CHAPTER 7

Estimate of the pressure

The pressure P satisfies

$$\begin{cases} -\Delta_{x,y} P = \partial_i\big(v^j \partial_j v^i\big) = \partial_i v^j \partial_j v^i \quad \text{in} \quad \Omega_t, \\ P|_{y=\eta} = 0, \quad \partial_y P|_{y=-1} = -1. \end{cases} \tag{7.1}$$

Here v is the velocity. In this section, we denote $P_1 \triangleq P + y$.

7.1. H^2 estimate of the pressure

LEMMA 7.1. *Let P be a solution of (7.1). It holds that*

$$\|\nabla_{x,y} P_1\|_{L^2(\Omega_t)} \leq K_\eta\big(\|\nabla_{x,y} v\|_{L^\infty(\Omega_t)}\|v\|_{L^2(\Omega_t)} + \|\eta\|_{H^{\frac12}}\big).$$

PROOF. Let $\overline{\eta}(t,x,y) = \chi(y)e^{(y-\eta(t,x))|D|}\eta$ for $y \leq \eta(t,x)$, where χ is a smooth function satisfying $\chi(0) = 1$ and $\chi(y) = 0$ for $y \leq -1 + h_0/2$. Then $P_2 \triangleq P + y - \overline{\eta}$ satisfies

$$\begin{cases} -\Delta_{x,y} P_2 = \Delta_{x,y}\overline{\eta} + \partial_i\big(v^j \partial_j v^i\big) \quad \text{in} \quad \Omega_t, \\ P_2|_{y=\eta} = 0, \quad \partial_y P_2|_{y=-1} = 0. \end{cases}$$

Thanks to $v^{d+1}|_{y=-1} = 0$, we get by integration by parts that

$$\begin{aligned} \int_{\Omega_t} |\nabla_{x,y} P_2|^2 dxdy &= \int_{\Omega_t} \big(\Delta_{x,y}\overline{\eta} + \partial_i\partial_j(v^j v^i)\big) P_2 dxdy \\ &= -\int_{\Omega_t} \big(\partial_j(v^j v^h) + \partial_y(v^h v^{d+1})\big) \cdot \nabla P_2 dxdy \\ &\quad - \int_{\Omega_t} \partial_y(v^{d+1}v^{d+1})\partial_y P_1 dxdy - \int_{\Omega_t} \nabla_{x,y}\overline{\eta} \cdot \nabla_{x,y} P_2 dxdy \\ &\leq C\big(\|\nabla_{x,y} v\|_{L^\infty(\Omega_t)}\|v\|_{L^2(\Omega_t)} + \|\nabla_{x,y}\overline{\eta}\|_{L^2(\Omega_t)}\big)\|\nabla_{x,y} P_2\|_{L^2(\Omega_t)}, \end{aligned}$$

from which, we deduce

$$\|\nabla_{x,y} P_2\|_{L^2(\Omega_t)} \leq C\big(\|\nabla_{x,y} v\|_{L^\infty(\Omega_t)}\|v\|_{L^2(\Omega_t)} + \|\nabla_{x,y}\overline{\eta}\|_{L^2(\Omega_t)}\big).$$

Then the lemma follows by uisng the fact that

$$\|\nabla_{x,y}\overline{\eta}\|_{L^2(\Omega_t)} \leq C(\|\eta\|_{W^{1,\infty}}, h_0)\|\eta\|_{H^{\frac12}}.$$

This completes the proof. □

Next, we give the higher estimates of pressure:

LEMMA 7.2. *Let P be a solution of (7.1). It holds that*

$$\|\nabla_{x,y} P_1\|_{H^1(\Omega_t)} \leq K_\eta\big(\|\nabla_{x,y} v\|_{L^\infty(\Omega_t)}\|v\|_{H^1(\Omega_t)} + \|\eta\|_{H^{\frac32}}\big).$$

PROOF. Let P_2 be as in the proof of Lemma 7.1. Then $\widetilde{P}_2$ satisfies

$$\left\{\begin{array}{l}\partial_z^2\widetilde{P}_2+\alpha\Delta\widetilde{P}_2+\beta\cdot\nabla\partial_z\widetilde{P}_2-\gamma\partial_z\widetilde{P}_2=F_0,\\ \widetilde{P}_2|_{z=0}=0,\quad \partial_z\widetilde{P}_2|_{z=-1}=0,\end{array}\right. \tag{7.2}$$

where $F_0=\alpha\widetilde{\partial_i v^j}\widetilde{\partial_j v^i}+\alpha\widetilde{\Delta_{x,y}\eta}$. It follows from (7.2) that

$$\begin{aligned}&\int_{\mathcal{S}}\partial_z^2\widetilde{P}_2\Delta\widetilde{P}_2dxdz+\int_{\mathcal{S}}\alpha|\Delta\widetilde{P}_2|^2dxdz+\int_{\mathcal{S}}\beta\cdot\nabla\partial_z\widetilde{P}_2\Delta\widetilde{P}_2dxdz\\&=\int_{\mathcal{S}}\gamma\partial_z\widetilde{P}_2\Delta\widetilde{P}_2dxdz+\int_{\mathcal{S}}F_0\Delta\widetilde{P}_2dxdz.\end{aligned}$$

By integration by parts, we get

$$\int_{\mathcal{S}}\partial_z^2\widetilde{P}_2\Delta\widetilde{P}_2dxdz=-\int_{\mathcal{S}}\partial_z^2\nabla\widetilde{P}_2\cdot\nabla\widetilde{P}_2dxdz=\int_{\mathcal{S}}|\nabla\partial_z\widetilde{P}_2|^2dxdz.$$

Thanks to the definition of α,β, it is easy to see that there exists $c>0$ depending on $\|\eta\|_{W^{1,\infty}},h_0$ so that

$$\begin{aligned}&\int_{\mathcal{S}}\partial_z^2\widetilde{P}_2\Delta\widetilde{P}_2dxdz+\int_{\mathcal{S}}\alpha|\Delta\widetilde{P}_2|^2dxdz+\int_{\mathcal{S}}\beta\cdot\nabla\partial_z\widetilde{P}_2\Delta\widetilde{P}_2dxdz\\&\geq c\int_{\mathcal{S}}\big(|\Delta\widetilde{P}_2|^2+|\nabla\partial_z\widetilde{P}_2|^2\big)dxdz.\end{aligned}$$

Hence, we obtain

$$\int_{\mathcal{S}}\big(|\Delta\widetilde{P}_1|^2+|\nabla\partial_z\widetilde{P}_1|^2\big)dxdz\leq C\big(\|\gamma\partial_z\widetilde{P}_1\|_{L^2(\mathcal{S})}+\|F_0\|_{L^2(\mathcal{S})}\big)\|\Delta\widetilde{P}_1\|_{L^2(\mathcal{S})}.$$

It follows from Lemma 4.4 that

$$\|\gamma\partial_z\widetilde{P}_2\|_{L^2(\mathcal{S})}\leq C\|\gamma\|_{L^2_z(I;L^\infty)}\|\partial_z\widetilde{P}_2\|_{L^\infty_z(I;L^2)}\leq K_\eta\|\partial_z\widetilde{P}_2\|_{L^2(\mathcal{S})}^{\frac12}\|\partial_z\nabla\widetilde{P}_1\|_{L^2(\mathcal{S})}^{\frac12}.$$

This shows that for any $\epsilon>0$,

$$\|\Delta\widetilde{P}_2\|_{L^2(\mathcal{S})}+\|\nabla\partial_z\widetilde{P}_2\|_{L^2(\mathcal{S})}\leq\|F_0\|_{L^2(\mathcal{S})}+K_\eta\|\partial_z\widetilde{P}_1\|_{L^2(\mathcal{S})}+\epsilon\|\partial_z\nabla\widetilde{P}_2\|_{L^2(\mathcal{S})}. \tag{7.3}$$

Using the equation (7.2), we infer that

$$\begin{aligned}\|\partial_z^2\widetilde{P}_2\|_{L^2(\mathcal{S})}\leq K_\eta\big(&\|\Delta\widetilde{P}_2\|_{L^2(\mathcal{S})}+\|\nabla\partial_z\widetilde{P}_2\|_{L^2(\mathcal{S})}\\&+\|\partial_z\widetilde{P}_2\|_{L^2(\mathcal{S})}^{\frac12}\|\partial_z^2\widetilde{P}_2\|_{L^2(\mathcal{S})}^{\frac12}+\|F_0\|_{L^2(\mathcal{S})}\big),\end{aligned}$$

which along with (7.3) gives by taking ϵ small that

$$\begin{aligned}\|\nabla_{x,z}^2\widetilde{P}_2\|_{L^2(\Omega_t)}\leq&K_\eta\big(\|F_0\|_{L^2(\mathcal{S})}+\|\partial_z\widetilde{P}_1\|_{L^2(\mathcal{S})}\big)\\ \leq&K_\eta\big(\|\eta\|_{H^{\frac32}}+\|\nabla_{x,y}v\|_{L^\infty(\Omega_t)}\|\nabla_{x,y}v\|_{L^2(\Omega_t)}+\|\partial_z\widetilde{P}_1\|_{L^2(\mathcal{S})}\big).\end{aligned}$$

Then the lemma follows by using Lemma 7.1. □

Next let us turn to H^1 estimate of $(\partial_t+v\cdot\nabla_{x,y})P\triangleq\mathcal{D}_tP$. Using the equation (1.1), a direct calculation gives

$$\left\{\begin{array}{l}\Delta_{x,y}\mathcal{D}_tP=\partial_kP\Delta_{x,y}v^k+G,\\ \mathcal{D}_tP|_{y=\eta}=0,\quad \partial_y\mathcal{D}_tP|_{y=-1}=\nabla\cdot v^h+\partial_yv^h\cdot\nabla P,\end{array}\right. \tag{7.4}$$

where

$$\begin{aligned}G =&4\delta^{ij}\partial_i v^k\partial_j\partial_k P-2(\partial_i v^j)(\partial_j v^k)\partial_k v^i\\ =&4\partial_k(\partial_i v^k\partial_i P)-2(\partial_i v^j)(\partial_j v^k)\partial_k v^i.\end{aligned}$$

Recall $\omega_{i,k}=\partial_i v^k-\partial_k v^i$. We have

$$\partial_k P\cdot\Delta_{x,y}v^k=\partial_i(\partial_k P\omega_{i,k})-\partial_i\partial_k P\omega_{i,k}=\partial_i(\partial_k P\omega_{i,k}).$$

LEMMA 7.3. *Let $\mathcal{D}_t P$ be a solution of (7.4). It holds that*

$$\|\nabla_{x,y}\mathcal{D}_t P\|_{L^2(\Omega_t)}\le K_\eta\big(1+\|\nabla_{x,y}v\|^2_{L^\infty(\Omega_t)}\big)\big(\|v\|_{H^1(\Omega_t)}+\|\eta\|_{H^{\frac32}}\big).$$

PROOF. We get by integration by parts that

$$\begin{aligned}&\int_{\Omega_t}|\nabla_{x,y}\mathcal{D}_t P|^2dxdy+\int_{\mathbf{R}^d}\partial_y\mathcal{D}_t P\mathcal{D}_t P|_{y=-1}dx\\ &=-\int_{\Omega_t}\big(4(\partial_i v^k\partial_i P)\partial_k\mathcal{D}_t P+\partial_k P\omega_{i,k}\partial_i\mathcal{D}_t P+2(\partial_i v^j)(\partial_j v^k)\partial_k v^i\mathcal{D}_t P\big)dxdy\\ &\quad+\int_{\mathbf{R}^d}(\partial_k P\omega_{d+1,k})\mathcal{D}_t P+(\partial_i v^{d+1}\partial_i P)\mathcal{D}_t Pdx\big|_{y=-1}\\ &\le C\big(\|\nabla_{x,y}v\|_{L^2(\Omega_t)}+\|\nabla_{x,y}v\|_{L^\infty(\Omega_t)}\|\nabla_{x,y}P_1\|_{L^2(\Omega_t)}\big)\|\nabla_{x,y}\mathcal{D}_t P\|_{L^2(\Omega_t)}\\ &\quad+C\|\nabla_{x,y}v\|^2_{L^\infty(\Omega_t)}\|\nabla_{x,y}v\|_{L^2(\Omega_t)}\|\mathcal{D}_t P\|_{L^2(\Omega_t)}\\ &\quad+C\|\nabla_{x,y}v\|_{L^\infty(\Omega_t)}\|\nabla_{x,y}P_1(\cdot,-1)\|_{L^2(\mathbf{R}^d)}\|\mathcal{D}_t P(\cdot,-1)\|_{L^2}\\ &\quad+\int_{\mathbf{R}^d}\nabla\cdot v^h\mathcal{D}_t P|_{y=-1}dx.\end{aligned}$$

Here we used $\operatorname{div} v=0$. By the boundary condition of (7.4), we have

$$\begin{aligned}&\int_{\mathbf{R}^d}\partial_y\mathcal{D}_t P\mathcal{D}_t P-\nabla\cdot v^h\mathcal{D}_t P|_{y=-1}dx\\ &\quad=\int_{\mathbf{R}^d}(\partial_y v^h\cdot\nabla P)\mathcal{D}_t P|_{y=-1}dx\\ &\quad\le\|\nabla_{x,y}v\|_{L^\infty(\Omega_t)}\|\nabla_{x,y}P_1(\cdot,-1)\|_{L^2}\|\mathcal{D}_t P(\cdot,-1)\|_{L^2}\\ &\quad\le K_\eta\|\nabla_{x,y}v\|_{L^\infty(\Omega_t)}\|\nabla_{x,y}P_1\|_{H^1(\Omega_t)}\|\mathcal{D}_t P\|_{H^1(\Omega_t)}\end{aligned}$$

Thanks to $\mathcal{D}_t P|_{y=\eta}=0$ so that

$$\|\mathcal{D}_t P\|_{L^2(\Omega_t)}\le K_\eta\|\mathcal{D}_t P\|_{H^1(\Omega_t)}.$$

we deduce that

$$\|\nabla_{x,y}\mathcal{D}_t P\|_{L^2(\Omega_t)}\le K_\eta\big(1+\|\nabla_{x,y}v\|^2_{L^\infty(\Omega_t)}\big)\big(\|\nabla_{x,y}P_1\|_{H^1(\Omega_t)}+\|\nabla_{x,y}v\|_{L^2(\Omega_t)}\big),$$

which together with Lemma 7.2 concludes the lemma. □

7.2. Hölder estimate of the pressure

In the sequel, we denote

(7.5) $$A(t)\triangleq 1+\|v(t)\|_{W^{1,\infty}(\Omega_t)}+\|v(t)\|_{H^1(\Omega_t)}+\|\eta(t)\|_{H^{\frac32}}.$$

First, we give the Hölder estimate of the pressure.

LEMMA 7.4. *Let $I_0=[-\frac34,0]$ and P be a solution of (7.1). Then it holds that*

$$\|\nabla_{x,z}\widetilde{P}\|_{\widetilde{L}^\infty_z(I_0;C^{\frac12})}\le K_\eta A(t)^2.$$

PROOF. Apply Proposition 4.16 with $F_0=\alpha\widetilde{\partial_i v^j}\widetilde{\partial_j v^i}$ and $G_0=0$ to obtain

$$\begin{aligned}\|\nabla_{x,z}\widetilde{P}\|_{\widetilde{L}^\infty_z(I_0;C^{\frac12})}\le& K_\eta\big(\|F_0\|_{\widetilde{L}^2_z(I;C^0)}+\|\nabla_{x,z}\widetilde{P}\|_{\widetilde{L}^\infty_z(I_0;C^{-\delta})}\\&+\|\nabla_{x,z}\widetilde{P}\|_{\widetilde{L}^2_z(I_1\setminus I_0;C^{-\frac{\varepsilon}{2}})}\big),\end{aligned}$$

where $I_1=[-\frac78,0]$ and $\delta>0$ is taken so that $-\delta+\frac d2\le-\frac12$. It is obvious that

$$\|F_0\|_{L^2_z(I;C^0)}\le K_\eta\|\nabla_{x,y}v\|^2_{L^\infty(\Omega_t)}.$$

It follows from Lemma 4.7 and Lemma 7.1 that

$$\begin{aligned}\|\nabla_{x,z}\widetilde{P}\|_{\widetilde{L}^\infty_z(I_0;C^{-\delta})}\le&\|\nabla_{x,z}\widetilde{P_1}\|_{\widetilde{L}^\infty_z(I_0;C^{-\delta})}+\|\nabla_{x,z}\rho_\delta\|_{\widetilde{L}^\infty_z(I_0;C^{-\delta})}\\\le& C\|\nabla_{x,z}\widetilde{P_1}\|_{L^\infty_z(I_0;H^{-\frac12})}+C\|\eta\|_{L^\infty}\\\le& K_\eta\big(\|\nabla_{x,y}P_1\|_{L^2(\Omega_t)}+\|F_0\|_{L^2_z(I;L^2)}\big)+C\|\eta\|_{L^\infty}\\\le& K_\eta\big(\|\nabla_{x,y}v\|_{L^\infty(\Omega_t)}\|v\|_{H^1(\Omega_t)}+\|\eta\|_{H^{\frac12}}+1\big).\end{aligned}$$

Take $c_1,c_2\in(0,1)$ depending on K_η,a such that $\rho_\delta(x,I_1\setminus I_0)\in\mathcal{S}_1=\big\{(x,y):y\in[-1+c_1h_0,\eta(x)-c_2h_0]\big\}$. Let δ_1 be as in Proposition 4.18. Then for p big enough depending on ε and d, we have

$$\begin{aligned}\|\nabla_{x,z}\widetilde{P}\|_{\widetilde{L}^2_z(I_1\setminus I_0;C^{-\frac{\varepsilon}{2}})}\le& C(\delta_1)\|\nabla_{x,z}\widetilde{P}_1\|_{L^2_z(I;L^2)}+C\|\eta\|_{W^{1,\infty}}\\&+\sup_{X_0\in\mathcal{S}_1}\|\nabla_{x,y}P\|_{L^p(B_{\delta_1}(X_0))},\end{aligned}$$

which together with Proposition 4.18 and Lemma 7.1 implies that

$$\begin{aligned}\|\nabla_{x,z}\widetilde{P}\|_{\widetilde{L}^2_z(I_1\setminus I_0;C^{-\frac{\varepsilon}{2}})}&\le K_\eta\big(\|\nabla_{x,y}v\|^2_{L^\infty(\Omega_t)}+\|\nabla_{x,y}P_1\|_{L^2(\Omega_t)}\big)+K_\eta\\&\le K_\eta A(t)^2.\end{aligned}\tag{7.6}$$

Putting the above estimates together concludes the lemma. $\square$

Next, we give the estimate of $\nabla_{x,z}\widetilde{\mathcal{D}_tP}$ in Besov space:

LEMMA 7.5. *Let $I_0=[-\frac34,0]$ and $\mathcal{D}_tP$ be a solution of (7.4). Then it holds that*

$$\|\nabla_{x,z}\widetilde{\mathcal{D}_tP}\|_{\widetilde{L}^\infty_z(I_0;B^0_{\infty,1})}\le K_\eta\big(1+\|\nabla_{x,z}\widetilde{v}\|_{L^\infty_z(I;B^0_{\infty,1})}\big)A(t)^2.$$

PROOF. We denote

$$\begin{aligned}F_0=&\sum_{i=1}^d\partial_i\big(\alpha\big(\widetilde{\partial_kP}\widetilde{\omega_{i,k}}+4\widetilde{\partial_kv^i}\widetilde{\partial_kP}\big)\big)-2\alpha(\widetilde{\partial_iv^j})(\widetilde{\partial_jv^k})\widetilde{\partial_kv^i}\\&+\Big(\sum_{i=1}^d(\partial_z\alpha_i-\partial_i\alpha)\big(\widetilde{\partial_kP}\widetilde{\omega_{i,k}}+4\widetilde{\partial_kv^i}\widetilde{\partial_kP}\big)\\&-\partial_z\alpha_{d+1}\big(\widetilde{\partial_kP}\widetilde{\omega_{d+1,k}}+4\widetilde{\partial_kv^{d+1}}\widetilde{\partial_kP}\big)\Big)\triangleq F_0^1+F_0^2+F_0^3,\end{aligned}\tag{7.7}$$

$$G_0=-\sum_{i=1}^d\alpha_i\big(\widetilde{\partial_kP}\widetilde{\omega_{i,k}}+4\widetilde{\partial_kv^i}\widetilde{\partial_kP}\big)+\alpha_{d+1}\big(\widetilde{\partial_kP}\widetilde{\omega_{d+1,k}}+4\widetilde{\partial_kv^{d+1}}\widetilde{\partial_kP}\big),\tag{7.8}$$

where $\alpha_1=\alpha\frac{\partial_i\rho_\delta}{\partial_z\rho_\delta}$ for $i=1,\cdots d$ and $\alpha_{d+1}=\frac{\alpha}{\partial_z\rho_\delta}$. Then $\mathcal{D}_tP$ satisfies

$$\partial_z^2\mathcal{D}_tP+\alpha\Delta\mathcal{D}_tP+\beta\cdot\nabla\partial_z\mathcal{D}_tP-\gamma\partial_z\mathcal{D}_tP=F_0+\partial_zG_0.$$

We apply Proposition 4.16 to obtain

$$\|\nabla_{x,z}\widetilde{\mathcal{D}_tP}\|_{\widetilde{L}^\infty_z(I_0;B^0_{\infty,1})} \le K_\eta\big(\|F_0\|_{Y^0_1(I)} + \|G_0\|_{\widetilde{L}^\infty_z(I_0;B^0_{\infty,1})} + \|\nabla_{x,z}\widetilde{\mathcal{D}_tP}\|_{\widetilde{L}^\infty_z(I_0;C^{-\delta})} + \|\nabla_{x,z}\widetilde{\mathcal{D}_tP}\|_{\widetilde{L}^2_z(I_1\setminus I_0;C^{-\frac{\varepsilon}{2}})}\big),$$

where $I_1=[-\frac{7}{8},0]$. By Lemma 7.6 and Lemma 7.4, we have

$$\begin{aligned}
&\|F_0\|_{Y^0_1(I)} + \|G_0\|_{\widetilde{L}^\infty_z(I_0;B^0_{\infty,1})}\\
&\le K_\eta\|\nabla_{x,z}\widetilde{v}\|_{\widetilde{L}^\infty_z(I;B^0_{\infty,1})}\big(\|\nabla_{x,y}v\|^2_{L^\infty(\Omega_t)} + \|\nabla_{x,z}\widetilde{P}\|_{\widetilde{L}^\infty_z(I;C^{\frac12})}\big)\\
&\le K_\eta\|\nabla_{x,z}\widetilde{v}\|_{\widetilde{L}^\infty_z(I;B^0_{\infty,1})}A(t)^2. \qquad (7.9)
\end{aligned}$$

We get by Lemma 7.3 and Lemmma 4.7 that

$$\begin{aligned}
\|\nabla\widetilde{\mathcal{D}_tP}\|_{\widetilde{L}^\infty_z(I_0;C^{-\delta})} &\le C\|\nabla\widetilde{\mathcal{D}_tP}\|_{L^\infty_z(I_0;H^{-\frac12})} \le C\|\nabla_{x,y}\mathcal{D}_tP\|_{L^2(\Omega_t)}\\
&\le K_\eta\big(1+\|\nabla_{x,y}v\|_{L^\infty(\Omega_t)}\big)A(t)^2.
\end{aligned}$$

On the other hand, for $z\in[-1,0]$,

$$\begin{aligned}
\partial_z\widetilde{\mathcal{D}_tP}(x,z) &= \int_{-1}^z \partial_z^2\widetilde{\mathcal{D}_tP}(x,z')dz'\\
&= G(z)-G(-1)+\int_{-1}^z\big(F_0-\alpha\Delta\widetilde{\mathcal{D}_tP}+\beta\nabla\partial_z\widetilde{\mathcal{D}_tP}-\gamma\partial_z\widetilde{\mathcal{D}_tP}\big)dz'.
\end{aligned}$$

From Lemma 4.8 in **[33]** and Lemma 7.3, we know that

$$\begin{aligned}
\|\alpha\Delta\widetilde{\mathcal{D}_tP}-\beta\nabla\partial_z\widetilde{\mathcal{D}_tP}+\gamma\partial_z\widetilde{\mathcal{D}_tP}\|_{L^1_z(-1,0;H^{-1})} &\le K_\eta\|\nabla_{x,z}\widetilde{\mathcal{D}_tP}\|_{L^2(\overline{\mathcal{S}})}\\
&\le K_\eta\big(1+\|\nabla_{x,y}v\|_{L^\infty(\Omega_t)}\big)A(t),
\end{aligned}$$

which along with (7.9) implies

$$\|\partial_z\widetilde{\mathcal{D}_tP}\|_{\widetilde{L}^\infty_z(I_0;C^{-\delta})} \le K_\eta\big(1+\|\nabla_{x,z}\widetilde{v}\|_{\widetilde{L}^\infty_z(I;B^0_{\infty,1})}\big)A(t)^2.$$

A similar argument leading to (7.6) yields

$$\begin{aligned}
&\|\nabla_{x,z}\widetilde{\mathcal{D}_tP}\|_{\widetilde{L}^2_z(I_1\setminus I_0;C^{-\frac{\varepsilon}{2}})}\\
&\le K_\eta\big(\|\nabla_{x,y}v\|_{L^\infty(\Omega_t)}\|\nabla_{x,y}P\|_{L^\infty(\Omega_t)} + \|\nabla_{x,y}v\|^3_{L^\infty(\Omega_t)} + \|\nabla_{x,y}\mathcal{D}_tP\|_{L^2(\Omega_t)}\big)\\
&\le K_\eta\big(1+\|\nabla_{x,y}v\|_{L^\infty(\Omega_t)}\big)A(t)^2.
\end{aligned}$$

Putting the above estimates together concludes the lemma. □

Next, we give the estimates of F_0 and G_0.

LEMMA 7.6. *Let F_0 and G_0 be given by (7.7) and (7.8) respectively. Then we have*

$$\begin{aligned}
&\|F_0\|_{Y^0_1(I)} + \|G_0\|_{\widetilde{L}^\infty_z(I;B^0_{\infty,1})}\\
&\le K_\eta\|\nabla_{x,z}\widetilde{v}\|_{\widetilde{L}^\infty_z(I;B^0_{\infty,1})}\big(\|\nabla_{x,y}v\|^2_{L^\infty(\Omega_t)} + \|\nabla_{x,z}\widetilde{P}\|_{L^\infty_z(I;C^{\frac12})}\big).
\end{aligned}$$

PROOF. We can deduce from Lemma 2.17, Lemma 2.18 that for $\delta<0$,

$$\begin{aligned}
\|fg\|_{\widetilde{L}^p_z(I;B^\delta_{\infty,1})} &\le C\|f\|_{\widetilde{L}^p_z(I;B^0_{\infty,1})}\|g\|_{L^\infty(\overline{\mathcal{S}})},\\
\|fg\|_{\widetilde{L}^p_z(I;B^0_{\infty,1})} &\le C\|f\|_{\widetilde{L}^p_z(I;B^0_{\infty,1})}\|g\|_{\widetilde{L}^\infty_z(I;B^0_{\infty,1})},
\end{aligned}$$

which together with Lemma 4.4 imply that

$$\begin{aligned}
&\|G_0\|_{\widetilde{L}^\infty_z(I;B^0_{\infty,1})} \le K_\eta \|\nabla_{x,y}v\|_{L^\infty(\Omega_t)} \|\nabla_{x,z}\widetilde{P}\|_{\widetilde{L}^\infty_z(I;C^{\frac12})},\\
&\|F_0^1\|_{\widetilde{L}^\infty_z(I;B^{-1}_{\infty,1})} \le K_\eta \|\nabla_{x,z}\widetilde{v}\|_{\widetilde{L}^\infty_z(I;B^0_{\infty,1})} \|\nabla_{x,z}\widetilde{P}\|_{\widetilde{L}^\infty_z(I;C^{\frac12})},\\
&\|F_0^2\|_{\widetilde{L}^2_z(I_0;B^{-\frac12}_{\infty,1})} \le K_\eta \|\nabla_{x,y}v\|^2_{L^\infty(\Omega_t)} \|\nabla_{x,z}\widetilde{v}\|_{\widetilde{L}^\infty_z(I;B^0_{\infty,1})},\\
&\|F_0^3\|_{\widetilde{L}^2_z(I;B^{-\frac12}_{\infty,1})} \le K_\eta \|\nabla_{x,y}v\|_{L^\infty(\Omega_t)} \|\nabla_{x,z}\widetilde{P}\|_{\widetilde{L}^\infty_z(I;C^{\frac12})}.
\end{aligned}$$

Here we also used the fact that

$$\|\widetilde{\nabla_{x,y}v}\|_{\widetilde{L}^\infty_z(I;B^0_{\infty,1})} \le K_\eta \|\nabla_{x,z}\widetilde{v}\|_{\widetilde{L}^\infty_z(I;B^0_{\infty,1})}.$$

This gives the lemma by the definition of $Y_1^0(I)$. □

7.3. Sobolev estimate of the pressure

LEMMA 7.7. *Let $I_0 = [-\frac12, 0]$ and P be a solution of (7.1). Then it holds that*

$$\begin{aligned}
&\|\nabla_{x,z}\widetilde{P_1}\|_{X^{s-\frac12}(I_0)} \le K_\eta A(t)^2 \big(\|\nabla_{x,z}\widetilde{v}\|_{L^2_z(I;H^{s-1})} + \|\eta\|_{H^{s+\frac12}}\big),\\
&\|\nabla_{x,y}P_1(\cdot,-1)\|_{H^s} \le K_\eta A(t)^2 \big(\|\nabla_{x,z}\widetilde{v}\|_{H^{s-\frac12}(\mathcal{S})} + \|\eta\|_{H^{s+\frac12}}\big).
\end{aligned}$$

PROOF. Apply Proposition 4.12 with $F_0 = \alpha(\widetilde{\partial_i v^j}\widetilde{\partial_j v^i})$ and $f = 0$ to obtain

$$\begin{aligned}
\|\nabla_{x,z}\widetilde{P_1}\|_{X^{s-\frac12}(I_0)} \le K_\eta \big(&\|\nabla_{x,z}\widetilde{P_1}\|_{L^2(\mathcal{S})} + \|F_0\|_{Y^{s-\frac12}(I)}\\
&+ \|\eta\|_{H^{s+\frac12}} \|\nabla_{x,z}\widetilde{P_1}\|_{L^\infty(\mathbf{R}^d\times I_1)}\big)
\end{aligned}$$

with $I_1 = [-\frac34, 0]$. By Lemma 2.11 and Lemma 4.4, we have

$$\begin{aligned}
\|F_0\|_{Y^{s-\frac12}(I)} \le& \|\widetilde{\partial_i v^j}\widetilde{\partial_j v^i}\|_{L^2_z(I;H^{s-1})} + \|(\alpha-1)(\widetilde{\partial_i v^j}\widetilde{\partial_j v^i})\|_{L^2_z(I;H^{s-1})}\\
\le& K_\eta \|\widetilde{\partial_i v^j}\|_{L^\infty(\mathcal{S})} \|\widetilde{\partial_j v^i}\|_{L^2_z(I;H^{s-1})} + C\|\widetilde{\partial_i v^j}\|^2_{L^\infty(\mathcal{S})} \|\alpha - 1\|_{L^2_z(I;H^{s-1})}\\
\le& K_\eta \big(\|\nabla_{x,y}v\|_{L^\infty(\Omega_t)} \|\nabla_{x,z}\widetilde{v}\|_{L^2_z(I;H^{s-1})} + \|\nabla_{x,y}v\|^2_{L^\infty(\Omega_t)} \|\eta\|_{H^{s-\frac12}}\big),
\end{aligned}$$

which along with Lemma 7.1 and Lemma 7.4 gives the first inequality.

In fact, the first inequality also holds with $[a, 0]$ for $a > -1$ instead of I_0. This implies there exists $y_0 \in (-1, -1 + h_0)$ so that

$$\|P_1(\cdot, y_0)\|_{H^s} \le K_\eta A(t)^2 \big(\|\nabla_{x,z}\widetilde{v}\|_{L^2_z(I;H^{s-1})} + \|\eta\|_{H^{s+\frac12}}\big).$$

Let $\mathcal{S}_1 = \mathbf{R}^d \times [-1, y_1]$ where $y_1 < y_0$. Then the standard elliptic estimate ensures that

$$\begin{aligned}
\|\nabla_{x,y}P_1\|_{H^{s+\frac12}(\mathcal{S}_1)} \le& C\big(\|P_1(\cdot,y_0)\|_{H^s} + \|\nabla_{x,y}v\nabla_{x,y}v\|_{H^{s-\frac12}(\mathcal{S}_1)}\big)\\
\le& K_\eta A(t)^2 \big(\|\nabla_{x,z}\widetilde{v}\|_{L^2_z(I;H^{s-1})} + \|\nabla_{x,y}v\|_{H^{s-\frac12}(\mathcal{S}_1)} + \|\eta\|_{H^{s+\frac12}}\big),
\end{aligned}$$

which together with the fact that

$$\|\nabla_{x,y}v\|_{H^{s-\frac12}(\mathcal{S}_1)} \le K_\eta \big(\|\nabla_{x,z}\widetilde{v}\|_{H^{s-\frac12}(\mathcal{S})} + \|\eta\|_{H^{s+\frac12}}\big),$$

implies the second inequality by the trace theorem. □

7.4. Estimate of a

Recall that $a(t,x) = -(\partial_y P)(t,x,\eta(t,x))$.

LEMMA 7.8. *It holds that*

$$\|a\|_{C^{\frac{1}{2}}} \le K_\eta A(t)^2,$$
$$\|D_t a\|_{B^0_{\infty,1}} \le K_\eta\big(1 + \|\widetilde{v}\|_{L^\infty_z(I;B^1_{\infty,1})}\big)A(t)^2.$$

PROOF. The first inequality is a direct consequence of Lemma 7.4 and Lemma 4.5 by using $\partial_y P = \frac{\partial_z \widetilde{P}}{\partial_z \rho_\delta}$. To show the second inequality, we write

$$\begin{aligned}\partial_t a + T_V\cdot\nabla a =& \partial_y(\partial_t + v\cdot\nabla_{x,y}P) - \partial_y v\cdot\nabla_{x,y}P|_{y=\eta} + (T_V - V)\cdot\nabla a\\ =& \frac{\partial_z\widetilde{\mathcal{D}_t P}}{\partial_z\rho_\delta} - \frac{\partial_z\widetilde{v}}{\partial_z\rho}\cdot\big(\nabla\widetilde{P} - \frac{\nabla\rho_\delta}{\partial_z\rho_\delta}\partial_z\widetilde{P}, \frac{\partial_z\widetilde{P}}{\partial_z\rho}\big)\big|_{z=0} + (T_V - V)\cdot\nabla a\\ \triangleq& I_1 + I_2 + I_3.\end{aligned}$$

Using Lemma 2.10, Lemma 2.12 and Lemma 4.5, it is easy to show that

$$\|I_1\|_{B^0_{\infty,1}} \le K_\eta\|\partial_z\widetilde{\mathcal{D}_t P}\|_{L^\infty_z(I;B^0_{\infty,1})},$$
$$\|I_2\|_{B^0_{\infty,1}} \le K_\eta\|\nabla_{x,z}\widetilde{P}\|_{L^\infty_z(I;C^{\frac{1}{2}})}\|\nabla_{x,z}\widetilde{v}\|_{L^\infty_z(I;B^0_{\infty,1})},$$
$$\|I_3\|_{B^0_{\infty,1}} \le C\|V\|_{B^1_{\infty,1}}\|a\|_{C^{\frac{1}{2}}} \le C\|\widetilde{v}\|_{L^\infty_z(I;B^1_{\infty,1})}\|a\|_{C^{\frac{1}{2}}},$$

which along with Lemma 7.4 and Lemma 7.5 lead to the second inequality. □

LEMMA 7.9. *It holds that*

$$\|a-1\|_{H^{s-\frac{1}{2}}} \le K_\eta A(t)^2\big(\|\nabla_{x,z}\widetilde{v}\|_{L^2_z(I;H^{s-1})} + \|\eta\|_{H^{s+\frac{1}{2}}}\big).$$

PROOF. Note that $a - 1 = \frac{1}{\partial_z\rho_\delta}\partial_z\widetilde{P_1}\big|_{z=0}$. Then we deduce from Lemma 2.11, Lemma 2.12 and Lemma 4.5 that

$$\|a-1\|_{H^{s-\frac{1}{2}}} \le K_\eta\big(\|\partial_z\widetilde{P_1}\|_{L^\infty_z([-\frac{1}{2},0];H^{s-\frac{1}{2}})} + \|\eta\|_{H^{s+\frac{1}{2}}}\big),$$

which together with Lemma 7.7 yields the result. □

CHAPTER 8

Estimate of the velocity

The velocity v satisfies

$$\begin{cases} \Delta_{x,y} v = -\nabla_{x,y} \times \omega \quad \text{in} \quad \Omega_t, \\ v|_{y=\eta} = (V, B), \quad v|_{y=-1} = (V_b, 0). \end{cases}$$

Here ω is the vorticity.

8.1. Sobolev estimate of the velocity

The following H^1 estimate is classical:

$$\|v\|_{H^1(\Omega_t)} \leq K_\eta \big(\|\omega\|_{L^2(\Omega_t)} + \|(V, B, V_b)\|_{H^{\frac{1}{2}}} \big). \tag{8.1}$$

This together with Proposition 4.12 and Proposition 4.14 ensures that

LEMMA 8.1. *Let $I_0 = [a, 0]$ for $a \in (-1, 0)$. It holds that*

$$\|\nabla_{x,z} \widetilde{v}\|_{X^{s-1}(I_0)} \leq K_\eta \big(\|(V, B, V_b)\|_{H^s} + \|\widetilde{\omega}\|_{H^{s-\frac{1}{2}}(\mathcal{S})} + \|\nabla_{x,y} v\|_{L^\infty(\Omega_t)} \|\eta\|_{H^{s+\frac{1}{2}}} \big).$$

If $s - \frac{1}{2}$ is an integer, we have

$$\|\nabla_{x,z} \widetilde{v}\|_{H^{s-\frac{1}{2}}(\mathcal{S})} \leq K_\eta \big(\|(V, B, V_b)\|_{H^s} + \|\widetilde{\omega}\|_{H^{s-\frac{1}{2}}(\mathcal{S})} + \|\nabla_{x,y} v\|_{L^\infty(\Omega_t)} \|\eta\|_{H^{s+\frac{1}{2}}} \big).$$

8.2. The estimate of the irrotational part

The irrotational part v_{ir} of the velocity is defined by

$$\begin{cases} \Delta_{x,y} v_{ir} = 0 \quad \text{in} \quad \Omega_t, \\ v_{ir}|_{y=\eta} = (V, B), \quad v_{ir}|_{y=-1} = 0. \end{cases}$$

LEMMA 8.2. *Let $I_0 = [-\frac{3}{4}, 0]$. It holds that*

$$\|\nabla_{x,z} \widetilde{v_{ir}}\|_{\widetilde{L}^\infty_z(I_0; C^0)} \leq K_\eta A(t),$$
$$\|\nabla_{x,z} \widetilde{v_{ir}}\|_{\widetilde{L}^\infty_z(I_0; B^0_{\infty,1})} \leq K_\eta \big(\|\widetilde{v}\|_{L^\infty(I; B^1_{\infty,1})} + \|v\|_{H^1(\Omega_t)} \big),$$

where $A(t)$ is given by (7.5).

PROOF. Let $I_1 = [-\frac{7}{8}, 0]$. It follows from Proposition 4.16 that

$$\begin{aligned} \|\nabla_{x,z} \widetilde{v_{ir}}\|_{\widetilde{L}^\infty_z(I_0; C^0)} \leq K_\eta \big(&\|(V, B)\|_{C^1} + \|\nabla_{x,z} \widetilde{v}_{ir}\|_{\widetilde{L}^\infty_z(I_0; C^{-\delta})} \\ &+ \|\nabla_{x,z} \widetilde{v}_{ir}\|_{\widetilde{L}^2_z(I_1 \setminus I_0; C^{-\frac{\epsilon}{2}})} \big) \end{aligned}$$

for any $\delta > 0$. Take $\delta > 0$ so that $-\delta + \frac{d}{2} \leq -\frac{1}{2}$. Then we get by Lemma 4.7 that

$$\begin{aligned} \|\nabla_{x,z} \widetilde{v_{ir}}\|_{\widetilde{L}^\infty_z(I_0; C^{-\delta})} \leq & C \|\nabla_{x,z} \widetilde{v}_{ir}\|_{L^\infty(I_0; H^{-\frac{1}{2}})} \\ \leq & K_\eta \|\nabla_{x,y} v_{ir}\|_{L^2(\Omega_t)} \leq K_\eta \|(V, B)\|_{H^{\frac{1}{2}}}. \end{aligned}$$

While, Proposition 4.18 implies that

$$\|\nabla_{x,z}\widetilde{v_{ir}}\|_{\widetilde{L}^2_z(I_1\setminus I_0;C^{-\frac{\epsilon}{2}})} \le K_\eta \|\nabla_{x,y}v_{ir}\|_{L^2(\Omega_t)} \le K_\eta\|(V,B)\|_{H^{\frac12}}.$$

This shows that

$$\|\nabla_{x,z}\widetilde{v_{ir}}\|_{\widetilde{L}^\infty_z(I_0;C^0)} \le K_\eta\big(\|(V,B)\|_{C^1}+\|(V,B)\|_{H^{\frac12}}\big) \le K_\eta A(t).$$

Similar argument leads to

$$\begin{aligned}\|\nabla_{x,z}\widetilde{v_{ir}}\|_{\widetilde{L}^\infty_z(I_0;B^0_{\infty,1})} \le& K_\eta\big(\|(V,B)\|_{B^1_{\infty,1}}+\|(V,B)\|_{H^{\frac12}}\big)\\ \le& K_\eta\big(\|\widetilde{v}\|_{L^\infty(I;B^1_{\infty,1})}+\|v\|_{H^1(\Omega_t)}\big).\end{aligned}$$

This finishes the proof. □

With Lemma 8.2, we deduce from Proposition 5.4 and Proposition 5.5 that

PROPOSITION 8.3. *It holds that*

$$\|R(\eta)(V,B)\|_{C^{\frac12}} \le K_\eta A(t),$$
$$\|R(\eta)(V,B)\|_{H^{s-1}} \le K_\eta\big(\|(V,B)\|_{H^{s-\frac12}}+A(t)\|\eta\|_{H^{s+\frac12}}\big),$$
$$\|R(\eta)(V,B)\|_{H^{s-\frac12}} \le K_\eta\big(\|(V,B)\|_{H^s}+(\|\widetilde{v}\|_{L^\infty(I;B^1_{\infty,1})}+A(t))\|\eta\|_{H^{s+\frac12}}\big).$$

8.3. The estimate of the rotational part

The rotational part v_ω of the velocity is defined by

$$\begin{cases}\Delta_{x,y}v_\omega = -\nabla_{x,y}\times\omega \quad \text{in} \quad \Omega_t,\\ v_\omega|_{y=\eta}=0, \quad v_\omega|_{y=-1}=(V_b,0).\end{cases}$$

LEMMA 8.4. *Let $I_0=[-\frac34,0]$. It holds that*

$$\|\nabla_{x,z}\widetilde{v_\omega}\|_{\widetilde{L}^\infty_z(I_0;C^0)} \le K_\eta A(t),$$
$$\|\nabla_{x,z}\widetilde{v_\omega}\|_{\widetilde{L}^\infty_z(I_0;B^0_{\infty,1})} \le K_\eta\big(\|\widetilde{v}\|_{L^\infty(I;B^1_{\infty,1})}+\|v\|_{H^1(\Omega_t)}\big),$$
$$\|\nabla_{x,z}\widetilde{v_\omega}\|_{X^{s-1}([-\frac12,0])} \le K_\eta\big(\|V_b\|_{H^{\frac12}}+\|\widetilde{\omega}\|_{H^{s-\frac12}(\mathcal{S})}+A(t)\|\eta\|_{H^s}\big).$$

PROOF. The first two inequalities follows from Lemma 8.2 and the fact $v_\omega = v-v_{ir}$. Then we get by Proposition 4.12 that

$$\begin{aligned}\|\nabla_{x,z}\widetilde{v_\omega}\|_{X^{s-1}([-\frac12,0])} \le K_\eta\big(&\|\nabla_{x,y}v_\omega\|_{L^2(\Omega_t)}+\|\widetilde{\omega}\|_{H^{s-\frac12}(\mathcal{S})}\\ &+\|\nabla_{x,y}\widetilde{v_\omega}\|_{\widetilde{L}^\infty_z(I_0;C^0)}\|\eta\|_{H^s}\big),\end{aligned}$$

which along with the first inequality yields the third inequality. □

Next we show that $(\partial_t+v\cdot\nabla_{x,y})v_\omega \triangleq \mathcal{D}_t v_\omega$ has the similar estimates. It is easy to verify that $\mathcal{D}_t v_\omega$ satisfies

$$\begin{cases}\Delta_{x,y}\mathcal{D}_t v_\omega = G_\omega \quad \text{in} \quad \Omega_t,\\ \mathcal{D}_t v_\omega|_{y=\eta}=0, \quad \mathcal{D}_t v_\omega|_{y=-1}=(\dot{V}_b,0).\end{cases}$$

where $\dot{V}_b=(\partial_t+V_b\cdot\nabla)V_b$ and

$$G_\omega = -\nabla_{x,y}\times\mathcal{D}_t\omega+\nabla_{x,y}\cdot(\omega\cdot\nabla_{x,y}\times v)+\nabla\times(\omega\cdot\nabla v_\omega)+2\partial_i(\nabla_k v^i\cdot\nabla_k v_\omega).$$

It is easy to find that G_ω can be rewritten $\widetilde{G}_\omega = \widetilde{G}^0_\omega + \nabla_z \widetilde{G}^1_\omega$ where $\widetilde{G}^0_\omega$ and $\widetilde{G}^1_\omega$ satisfy that (by Lemma 2.13 and Lemma 4.3),

$$\begin{aligned}&\|\widetilde{G}^0_\omega\|_{L^2(I_0;H^{s-\frac{3}{2}})} + \|\widetilde{G}^1_\omega\|_{L^2(I_0;H^{s-\frac{1}{2}})}\\ &\le K_\eta\big(\|\nabla_{x,y}v\|^2_{L^\infty(\Omega_t)} + \|\nabla_{x,y}v\|_{L^\infty(\Omega_t)}\|\nabla_{x,y}v_\omega\|_{L^\infty(\Omega_t)}\big)\|\eta\|_{H^{s+\frac{1}{2}}}\\ &+ K_\eta\big(\|\nabla_{x,y}v\|_{L^\infty(\Omega_t)} + \|\nabla_{x,y}v_\omega\|_{L^\infty(\Omega_t)}\big)\|\nabla_{x,z}\widetilde{v}\|_{L^2(I_0;H^{s-\frac{1}{2}})}\\ &+ K_\eta\|\nabla_{x,y}v\|_{L^\infty(\Omega_t)}\|\nabla_{x,z}\widetilde{v_\omega}\|_{L^2(I_0;H^{s-\frac{1}{2}})}.\end{aligned}$$

LEMMA 8.5. *It holds that*

$$\begin{aligned}&\|\nabla_{x,z}\widetilde{\mathcal{D}_t v_\omega}\|_{\widetilde{L}^\infty([-\frac{1}{2},0];C^0)} \le K_\eta\big(1 + \|\widetilde{v}\|_{L^\infty(I;B^1_{\infty,1})}\big)A(t)^2,\\ &\|\nabla_{x,z}\widetilde{\mathcal{D}_t v_\omega}\|_{X^{s-1}([-\frac{1}{2},0])} \le K_\eta\big(1 + \|\widetilde{v}\|_{L^\infty(I;B^1_{\infty,1})}\big)A(t)^2\\ &\qquad\qquad\times\big(\|\eta\|_{H^{s+\frac{1}{2}}} + \|(V,B,V_b)\|_{H^s} + \|\widetilde{\omega}\|_{H^{s-\frac{1}{2}}}\big).\end{aligned}$$

PROOF. Thanks to $\mathcal{D}_t\omega = \omega\cdot\nabla v$, we get

$$\|\mathcal{D}_t\omega\|_{L^2(\Omega_t)} \le \|\nabla_{x,y}v\|_{L^\infty(\Omega_t)}\|\nabla_{x,y}v\|_{L^2(\Omega_t)},$$

and by (6.3) and Lemma 7.2, we have

$$\|\dot{V}_b\|_{H^{\frac{1}{2}}} \le K_\eta\|\nabla_{x,y}P\|_{H^1(\Omega_t)} \le K_\eta A(t)^2.$$

Then it is easy to show that

$$\begin{aligned}\|\nabla_{x,y}\mathcal{D}_t v_\omega\|_{L^2(\Omega_t)} \le& K_\eta\big(\|\dot{V}_b\|_{H^{\frac{1}{2}}} + \|\nabla_{x,y}v\|_{L^\infty(\Omega_t)}(\|\nabla_{x,y}v\|_{L^2(\Omega_t)} + \|\nabla_{x,y}v_\omega\|_{L^2(\Omega_t)})\big)\\ \le& K_\eta A(t)^2.\end{aligned}$$

Then a similar argument leading to Lemma 7.5 ensures that

$$\begin{aligned}\|\nabla_{x,z}\widetilde{\mathcal{D}_t v_\omega}\|_{L^\infty([-\frac{3}{4},0];C^0)} \le& K_\eta\big(A(t)^2 + \|\nabla_{x,y}v_\omega\|_{L^\infty(\Omega_t)}A(t)\big)\\ \le& K_\eta\big(1 + \|\widetilde{v}\|_{L^\infty(I;B^1_{\infty,1})}\big)A(t)^2.\end{aligned}$$

Let $I_0 = [-\frac{1}{2},0]$ and $I_1 = [-\frac{3}{4},0]$. Then Proposition 4.12 together with Lemma 8.1 and Lemma 8.4 implies that

$$\begin{aligned}\|\nabla_{x,z}\widetilde{\mathcal{D}_t v_\omega}\|_{X^{s-1}(I_0)} \le& K_\eta\big(\|\nabla_{x,y}\mathcal{D}_t v_\omega\|_{L^2(\Omega_t)} + \|G^0_\omega\|_{L^2(I_0;H^{s-\frac{3}{2}})} + \|G^1_\omega\|_{L^2(I_0;H^{s-\frac{1}{2}})}\\ &+ \|\nabla_{x,z}\widetilde{\mathcal{D}_t v_\omega}\|_{L^\infty(I_1;C^0)}\|\eta\|_{H^{s+\frac{1}{2}}}\big)\\ \le& K_\eta\big(1 + \|\widetilde{v}\|_{L^\infty(I;B^1_{\infty,1})}\big)A(t)^2\\ &\times\big(\|\eta\|_{H^{s+\frac{1}{2}}} + \|(V,B,V_b)\|_{H^s} + \|\widetilde{\omega}\|_{H^{s-\frac{1}{2}}}\big).\end{aligned}$$

The proof is finished. □

CHAPTER 9

Proof of break-down criterion

In this chapter, we assume that (η, v) is a solution of the system (1.1)–(1.5) obtained in Theorem 1.1 in $\Omega_t = \{(x,y) \in \mathbf{R}^d \times \mathbf{R} : -1 < y < \eta(t,x)\}$ for $t \in [0,T]$. We denote

$$\begin{aligned} A(t) &\triangleq 1 + \|v(t)\|_{W^{1,\infty}(\Omega_t)} + \|v(t)\|_{H^1(\Omega_t)} + \|\eta(t)\|_{H^{\frac{3}{2}}} + \frac{1}{c_0}, \\ \mathcal{B}(t) &\triangleq 1 + \|\widetilde{v}\|_{L^\infty_z(I; B^1_{\infty,1})}. \end{aligned}$$

9.1. The H^1 energy estimate

We have the following basic energy law for the system (1.1)–(1.5).

LEMMA 9.1. *For any $t \in [0,T]$, there holds*

$$E(t) = E(0), \quad E(t) \triangleq \|v(t)\|^2_{L^2(\Omega_t)} + \|\eta(t)\|^2_{L^2}.$$

PROOF. By (1.1), (1.4) and integration by parts, we get

$$\begin{aligned} \frac{d}{dt}\int_{\Omega_t} |v(t,x,y)|^2 dxdy &= \int_{\mathbf{R}^d} \partial_t\eta |v|^2 dx + 2\int_{\Omega_t} \partial_t v \cdot v dxdy \\ &= \int_{\mathbf{R}^d} \partial_t\eta |v|^2 dx - 2\int_{\Omega_t} \big(v\cdot\nabla_{x,y} v + \nabla(P+y)\big)\cdot v dxdy \\ &= \int_{\mathbf{R}^d} \partial_t\eta |v|^2 dx - \int_{\mathbf{R}^d} v\cdot \mathbf{n}_+ (|v|^2 + 2\eta)\sqrt{1+|\nabla\eta|^2} dx \\ &= -2\int_{\mathbf{R}^d} \partial_t\eta(t,x)\eta(t,x) dx = -\frac{d}{dt}\int_{\mathbf{R}^d} |\eta(t,x)|^2 dx. \end{aligned}$$

This shows that

$$\frac{d}{dt}\Big(\int_{\Omega_t} |v(t,x,y)|^2 dxdy + \int_{\mathbf{R}^d} |\eta(t,x)|^2 dx\Big) = 0.$$

Hence, $E(t) = E(0)$ for $t \in [0,T]$. □

LEMMA 9.2. *It holds that for any $t \in [0,T]$,*

$$\frac{d}{dt}\|\nabla_{x,y} v(t)\|_{L^2(\Omega_t)} \le K_\eta\big(\|v(t)\|_{W^{1,\infty}(\Omega_t)}\|v(t)\|_{H^1(\Omega_t)} + \|\eta(t)\|_{H^{\frac{3}{2}}}\big).$$

PROOF. Similar to the proof of Lemma 9.1, we have

$$\begin{aligned}
&\frac{d}{dt}\int_{\Omega_t}|\nabla_{x,y}v(t,x,y)|^2dxdy\\
&=\int_{\mathbf{R}^d}\partial_t\eta|\nabla_{x,y}v|^2dx+2\int_{\Omega_t}\partial_t\nabla_{x,y}v\cdot\nabla_{x,y}vdxdy\\
&=\int_{\mathbf{R}^d}\partial_t\eta|\nabla_{x,y}v|^2dx-\int_{\Omega_t}v\cdot\nabla_{x,y}|\nabla_{x,y}v|^2dxdy\\
&\quad-2\int_{\Omega_t}\big(\nabla_{x,y}v\cdot\nabla_{x,y}v+\nabla^2_{x,y}(P+y)\big)\cdot\nabla_{x,y}vdxdy\\
&=-2\int_{\Omega_t}\big(\nabla_{x,y}v\cdot\nabla_{x,y}v+\nabla^2_{x,y}(P+y)\big)\cdot\nabla_{x,y}vdxdy\\
&\le 2\|\nabla_{x,y}v\|_{L^\infty(\Omega_t)}\|\nabla_{x,y}v\|^2_{L^2(\Omega_t)}+\|\nabla^2_{x,y}(P+y)\|_{L^2(\Omega_t)}\|\nabla_{x,y}v\|_{L^2(\Omega_t)},
\end{aligned}$$

which along with Lemma 7.2 yields the result. □

9.2. Energy estimate of the trace of the velocity and the free surface

Let us first present the lower order energy estimate.

PROPOSITION 9.3. *It holds that*

$$\begin{aligned}
&\frac{d}{dt}\big(\|(V,B)\|^2_{H^{s-\frac12}}+\|V_b\|^2_{H^s}+\|\eta\|^2_{H^s}\big)\\
&\le K_\eta A(t)^2\Big(\|(V,B,V_b)\|^2_{H^s}+\|\nabla_{x,z}\widetilde{v}\|^2_{H^{s-\frac12}(\mathcal{S})}+\|\eta\|^2_{H^{s+\frac12}}\Big).
\end{aligned}$$

PROOF. Recall that $\eta(t,x)$ satisfies

$$\partial_t\eta+V\cdot\nabla\eta=B.$$

Make H^s energy estimate to obtain

$$\frac12\frac{d}{dt}\|\eta(t)\|^2_{H^s}\le-\big\langle\langle D\rangle^s(V\cdot\nabla\eta),\langle D\rangle^s\eta\big\rangle+\|B\|_{H^s}\|\eta\|_{H^s}.$$

We write

$$\begin{aligned}
\big\langle\langle D\rangle^s(V\cdot\nabla\eta),\langle D\rangle^s\eta\big\rangle=&\big\langle\langle D\rangle^s(T_V\cdot\nabla\eta),\langle D\rangle^s\eta\big\rangle+\big\langle\langle D\rangle^s((V-T_V)\cdot\nabla\eta),\langle D\rangle^s\eta\big\rangle\\
=&\big\langle[\langle D\rangle^s,T_V]\cdot\nabla\eta,\langle D\rangle^s\eta\big\rangle-\big\langle T_{\nabla\cdot V}\langle D\rangle^s\eta,\langle D\rangle^s\eta\big\rangle\\
&+\big\langle\langle D\rangle^s((V-T_V)\cdot\nabla\eta),\langle D\rangle^s\eta\big\rangle.
\end{aligned}$$

Then we deduce from Lemma 2.19 and Lemma 2.10 that

$$\big\langle\langle D\rangle^s(V\cdot\nabla\eta),\langle D\rangle^s\eta\big\rangle\le C\|V\|_{W^{1,\infty}}\|\eta\|^2_{H^s}+K_\eta\|V\|_{H^s}\|\eta\|_{H^s}.$$

This shows that

$$\frac{d}{dt}\|\eta(t)\|^2_{H^s}\le K_\eta A(t)\big(\|(V,B)\|^2_{H^s}+\|\eta\|^2_{H^s}\big).\tag{9.1}$$

Recall that (V,B,V_b) satisfies

$$\begin{aligned}
&(\partial_t+V\cdot\nabla)B=a-1,\\
&(\partial_t+V\cdot\nabla)V+a\zeta=0,\\
&(\partial_t+V_b\cdot\nabla)V_b+\nabla P|_{y=-1}=0.
\end{aligned}$$

In a similar way leading to (9.1), we deduce

$$
\begin{aligned}
&\frac{1}{2}\frac{d}{dt}\big(\|(V,B)\|^2_{H^{s-\frac12}}+\|V_b\|^2_{H^s}\big)\\
&\le C\|(\nabla V,\nabla B)\|_{L^\infty}\|(V,B)\|^2_{H^{s-\frac12}}+\|a-1\|_{H^{s-\frac12}}\|B\|_{H^{s-\frac12}}\\
&\qquad+\|a\zeta\|_{H^{s-\frac12}}\|V\|_{H^{s-\frac12}}+C\|\nabla V_b\|_{L^\infty}\|V_b\|^2_{H^s}+\|\nabla P|_{y=-1}\|_{H^s}\|V_b\|_{H^s},
\end{aligned}
$$

which together with Lemma 7.9 and Lemma 7.7 yields

$$
\begin{aligned}
&\frac{d}{dt}\big(\|(V,B)\|^2_{H^{s-\frac12}}+\|V_b\|^2_{H^s}\big)\\
&\le K_\eta A(t)^2\big(\|(V,B,V_b)\|^2_{H^s}+\|\nabla_{x,z}\widetilde{v}\|^2_{H^{s-\frac12}(\mathcal{S})}+\|\eta\|^2_{H^{s+\frac12}}\big),
\end{aligned}
$$

from which and (9.1), we deduce the proposition. □

Next we present the high order energy estimate.

PROPOSITION 9.4. *Let U be a solution of (6.13). Then there holds*

$$
\begin{aligned}
&\frac{1}{2}\frac{d}{dt}\big(\|D_tU\|^2_{H^{s-\frac12}}+\|T_{\sqrt{a\lambda}}U\|^2_{H^{s-\frac12}}\big)\\
&\le \big\langle (f+f_\omega)_{s-1/2},(D_tU)_{s-1/2}\big\rangle+K_\eta\mathcal{B}(t)A(t)^2\big(\|D_tU\|^2_{H^{s-\frac12}}+\|U\|^2_{H^s}\big).
\end{aligned}
$$

Here we denote $f_s\triangleq\langle D\rangle^s f$.

PROOF. It follows from (6.13) that $(D_tU)_{s-1/2}$ satisfies

$$
\begin{aligned}
D_t(D_tU)_{s-1/2}+T_{a\lambda}U_{s-1/2}=&[D_t,\langle D\rangle^{s-1/2}]D_tU+[T_{a\lambda},\langle D\rangle^{s-1/2}]U\\
&+(f+f_\omega)_{s-1/2}\triangleq F.
\end{aligned}
$$

Taking L^2 inner product with $(D_tU)_{s-1/2}$, we get

$$
\big\langle D_t(D_tU)_{s-1/2},(D_tU)_{s-1/2}\big\rangle+\big\langle T_{a\lambda}U_{s-1/2},(D_tU)_{s-1/2}\big\rangle=\big\langle F,(D_tU)_{s-1/2}\big\rangle. \tag{9.2}
$$

By integration by parts, we get

$$
\begin{aligned}
&\big\langle D_t(D_tU)_{s-1/2},(D_tU)_{s-1/2}\big\rangle\\
&=\frac{1}{2}\frac{d}{dt}\big\langle (D_tU)_{s-1/2},(D_tU)_{s-1/2}\big\rangle-\frac12\big\langle\nabla\cdot V(D_tU)_{s-1/2},(D_tU)_{s-1/2}\big\rangle\\
&\quad+\big\langle (T_V\cdot\nabla-V\cdot\nabla)(D_tU)_{s-1/2},(D_tU)_{s-1/2}\big\rangle\\
&\ge\frac{1}{2}\frac{d}{dt}\big\langle (D_tU)_{s-1/2},(D_tU)_{s-1/2}\big\rangle-C\|V\|_{B^1_{\infty,1}}\|D_tU\|^2_{H^{s-\frac12}},
\end{aligned}
\tag{9.3}
$$

where we used the inequality

$$
\|(T_V\cdot\nabla-V\cdot\nabla)(D_tU)_{s-1/2}\|_{L^2}\le C\|V\|_{B^1_{\infty,1}}\|D_tU\|_{H^{s-\frac12}}.
$$

Similarly, we have

$$\begin{aligned}
&\big\langle T_{a\lambda}U_{s-1/2},(D_tU)_{s-1/2}\big\rangle\\
&=\big\langle (T_{\sqrt{a\lambda}})^*T_{\sqrt{a\lambda}}U_{s-1/2},(D_tU)_{s-1/2}\big\rangle\\
&\quad+\big\langle \big(T_{a\lambda}-(T_{\sqrt{a\lambda}})^*T_{\sqrt{a\lambda}}\big)U_{s-1/2},(D_tU)_{s-1/2}\big\rangle\\
&=\big\langle (T_{\sqrt{a\lambda}}U_{s-1/2},T_{\sqrt{a\lambda}}(D_tU)_{s-1/2}\big\rangle+\big\langle \big(T_{a\lambda}-(T_{\sqrt{a\lambda}})^*T_{\sqrt{a\lambda}}\big)U_{s-1/2},(D_tU)_{s-1/2}\big\rangle\\
&=\frac{1}{2}\frac{d}{dt}\langle T_{\sqrt{a\lambda}}U_{s-1/2},T_{\sqrt{a\lambda}}U_{s-1/2}\rangle-\frac{1}{2}\langle\nabla\cdot VT_{\sqrt{a\lambda}}U_{s-1/2},T_{\sqrt{a\lambda}}U_{s-1/2}\rangle\\
&\quad+\big\langle T_{\sqrt{a\lambda}}U_{s-1/2},[T_{\sqrt{a\lambda}}\langle D\rangle^{s-\frac{1}{2}},D_t]U\big\rangle\\
&\quad+\big\langle T_{\sqrt{a\lambda}}U_{s-1/2},(T_V\cdot\nabla-V\cdot\nabla)T_{\sqrt{a\lambda}}U_{s-1/2}\big\rangle\\
&\quad+\big\langle \big(T_{a\lambda}-(T_{\sqrt{a\lambda}})^*T_{\sqrt{a\lambda}}\big)U_{s-1/2},(D_tU)_{s-1/2}\big\rangle.
\end{aligned}$$

It follows from Proposition 2.6 that

$$\begin{aligned}
&\big\langle\nabla\cdot VT_{\sqrt{a\lambda}}U_{s-1/2},T_{\sqrt{a\lambda}}U_{s-1/2}\big\rangle\le C\|\nabla V\|_{L^\infty}M_0^{\frac{1}{2}}(\sqrt{a\lambda})^2\|U\|_{H^s}^2,\\
&\big\langle\big(T_{a\lambda}-(T_{\sqrt{a\lambda}})^*T_{\sqrt{a\lambda}}\big)U_{s-1/2},(D_tU)_{s-1/2}\big\rangle\le CM_0^{\frac{1}{2}}(\sqrt{a\lambda})^2\|U\|_{H^s}\|D_tU\|_{H^{s-\frac{1}{2}}},\\
&\big\langle T_{\sqrt{a\lambda}}U_{s-1/2},(T_V\cdot\nabla-V\cdot\nabla)T_{\sqrt{a\lambda}}U_{s-1/2}\big\rangle\le M_0^{\frac{1}{2}}(\sqrt{a\lambda})^2\|V\|_{B^1_{\infty,\infty}}\|D_tU\|_{H^{s-\frac{1}{2}}}^2.
\end{aligned}$$

We get by Proposition 2.21 that

$$\begin{aligned}
&\big\langle T_{\sqrt{a\lambda}}U_{s-1/2},[T_{\sqrt{a\lambda}}\langle D\rangle^{s-\frac{1}{2}},D_t]U\big\rangle\\
&\le CM_0^{\frac{1}{2}}(\sqrt{a\lambda})\big(M_0^{\frac{1}{2}}(\sqrt{a\lambda})\|V\|_{B^1_{\infty,\infty}}+M_0^{\frac{1}{2}}(D_t\sqrt{a\lambda})\big)\|U\|_{H^s}^2.
\end{aligned}$$

This proves that

$$\begin{aligned}
\big\langle T_{a\lambda}U_{s-1/2},(D_tU)_{s-1/2}\big\rangle&\ge\frac{1}{2}\frac{d}{dt}\big\langle T_{\sqrt{a\lambda}}U_{s-1/2},T_{\sqrt{a\lambda}}U_{s-1/2}\big\rangle\\
&-C\|V\|_{B^1_{\infty,\infty}}M_0^{\frac{1}{2}}(\sqrt{a\lambda})^2\big(\|U\|_{H^s}^2+\|D_tU\|_{H^{s-\frac{1}{2}}}^2\big)-M_0^{\frac{1}{2}}(D_t\sqrt{a\lambda})\|U\|_{H^s}^2\\
&-CM_0^{\frac{1}{2}}(\sqrt{a\lambda})^2\|U\|_{H^s}\|D_tU\|_{H^{s-\frac{1}{2}}}.
\end{aligned}\tag{9.4}$$

By Lemma 2.19 and Proposition 2.6, we have

$$\|[D_t,\langle D\rangle^{s-1/2}]D_tU\|_{L^2}\le C\|V\|_{W^{1,\infty}}\|D_tU\|_{H^{s-\frac{1}{2}}},\tag{9.5}$$

$$\|[T_{a\lambda},\langle D\rangle^{s-1/2}]U\|_{L^2}\le CM_{\frac{1}{2}}^1(a\lambda)\|U\|_{H^s}.\tag{9.6}$$

Then it follows from (9.2)-(9.6) that

$$\begin{aligned}
&\frac{1}{2}\frac{d}{dt}\big(\|D_tU\|_{H^{s-\frac{1}{2}}}^2+\|T_{\sqrt{a\lambda}}U\|_{H^{s-\frac{1}{2}}}^2\big)\le\langle (f+f_\omega)_{s-1/2},(D_tU)_{s-1/2}\rangle\\
&\quad+C\big(1+\|V\|_{B^1_{\infty,\infty}}\big)\big(1+M_0^{\frac{1}{2}}(\sqrt{a\lambda})^2+M_{\frac{1}{2}}^1(a\lambda)\big)\big(\|D_tU\|_{H^{s-\frac{1}{2}}}^2+\|U\|_{H^s}^2\big)\\
&\quad+CM_0^{\frac{1}{2}}(D_t\sqrt{a\lambda})\|U\|_{H^s}^2,
\end{aligned}$$

while, by Lemma 7.8 and Lemma 4.6, we have

$$\begin{aligned}
M_0^{\frac{1}{2}}(\sqrt{a\lambda})^2+M_{\frac{1}{2}}^1(a\lambda)&\le K_\eta A(t)^2,\\
M_0^{\frac{1}{2}}(D_t\sqrt{a\lambda})&\le K_\eta\mathcal{B}(t)A(t)^2.
\end{aligned}$$

Then the proposition follows easily. □

9.3. Energy estimate of the vorticity

Using the equation (6.4), it is easy to see that the vorticity $\widetilde{\omega}(t,x,z)$ satisfies

$$\partial_t\widetilde{\omega}+\overline{v}\cdot\nabla_{x,z}\widetilde{\omega}=\widetilde{\omega}^h\cdot\big(\nabla\widetilde{v}-\frac{\nabla\rho_\delta}{\partial_z\rho_\delta}\partial_z\widetilde{v}\big)+\widetilde{\omega}^{d+1}\frac{\partial_z\widetilde{v}}{\partial_z\rho_\delta}\triangleq F,$$

where $\widetilde{v}^h=\big(\widetilde{v}^1,\cdots,\widetilde{v}^d\big),\widetilde{\omega}^h=\big(\widetilde{\omega}^1,\cdots,\widetilde{\omega}^d\big)$ and

$$\overline{v}=\big(\widetilde{v}^h,\frac{1}{\partial_z\rho_\delta}(\widetilde{v}^{d+1}-\partial_t\rho_\delta-\widetilde{v}^h\cdot\nabla\rho_\delta)\big).$$

PROPOSITION 9.5. *Let $s>\frac{d}{2}+1$ and $s-\frac12$ be an integer. Then we have*

$$\frac{d}{dt}\|\widetilde{\omega}(t)\|^2_{H^{s-\frac12}(\mathcal{S})}\le K_\eta\mathcal{B}(t)A(t)\big(\|\widetilde{\omega}\|^2_{H^{s-\frac12}(\mathcal{S})}+\|\eta\|^2_{H^s}+\|\widetilde{v}\|^2_{H^{s+\frac12}(\mathcal{S})}\big).$$

We need the following lemma.

LEMMA 9.6. *Let $s>1+\frac{d}{2}$. Then we have*

$$\|\nabla_{x,z}\overline{v}\|_{H^{s-\frac12}(\mathcal{S})}\le K_\eta\big(\|\widetilde{v}\|_{H^{s+\frac12}(\mathcal{S})}+\mathcal{B}(t)\|\eta\|_{H^s}\big),$$
$$\|\nabla_{x,z}\overline{v}\|_{L^\infty(\mathcal{S})}\le K_\eta\mathcal{B}(t).$$

PROOF. Thanks to the definition of $\overline{v}$, it suffices to consider $\overline{v}^{d+1}$. Let $\phi=\widetilde{v}^{d+1}-\partial_t\rho_\delta-\widetilde{v}^h\cdot\nabla\rho_\delta$. Let $\Delta^\delta_{x,z}=\Delta+\delta^{-2}\partial_z^2$ and $\nabla^\delta_{x,z}=(\nabla,\delta^{-1}\partial_z)$. Using the fact that

$$\Delta^\delta_{x,z}\rho_\delta=2\delta^{-1}e^{\delta z|D|}|D|\eta\triangleq f_\eta,$$

we find that

$$\Delta^\delta_{x,z}\phi=\Delta^\delta_{x,z}\widetilde{v}^{d+1}-\partial_tf_\eta-\Delta^\delta_{x,z}\widetilde{v}^h\cdot\nabla\rho_\delta-2\nabla^\delta_{x,z}\widetilde{v}^h\cdot\nabla\nabla^\delta_{x,z}\rho_\delta-\widetilde{v}\cdot\nabla f_\eta\triangleq F_{\eta,v}$$

together with the boundary condition $\phi=0$ on $z=0$ and $z=-1$.

By (1.4), we have

$$\|\partial_tf_\eta\|_{H^{s-\frac32}(\mathcal{S})}\le C\|\partial_t\eta\|_{H^{s-1}}\le C\big(\|\nabla\eta\|_{L^\infty}\|V\|_{H^{s-1}}+\|V\|_{L^\infty}\|\eta\|_{H^s}+\|B\|_{H^{s-1}}\big),$$

which together with Lemma 2.13, Lemma 4.3 implies that

$$\|F_{\eta,v}\|_{H^{s-\frac32}(\mathcal{S})}\le K_\eta\big(\|\widetilde{v}\|_{H^{s+\frac12}(\mathcal{S})}+\|\widetilde{v}\|_{W^{1,\infty}(\mathcal{S})}\|\eta\|_{H^s}\big).$$

Then Proposition 4.1 ensures that

$$\|\nabla_{x,z}\phi\|_{H^{s-\frac12}(\mathcal{S})}\le K_\eta\big(\|\widetilde{v}\|_{H^{s+\frac12}(\mathcal{S})}+\|\widetilde{v}\|_{W^{1,\infty}(\mathcal{S})}\|\eta\|_{H^s}\big).\tag{9.7}$$

We write

$$\begin{aligned}F_{\eta,v}=&\nabla^\delta_{x,z}\cdot\big(\nabla^\delta_{x,z}\widetilde{v}_{d+1}-\nabla^\delta_{x,z}\widetilde{v}^h\cdot\nabla\rho_\delta+(\widetilde{v}^h,0)f_\eta\big)\\&+\big(-\partial_tf_\eta-\nabla^\delta_{x,z}\widetilde{v}^h\cdot\nabla^\delta_{x,z}\nabla\rho_\delta+\nabla\cdot\widetilde{v}^hf_\eta\big)\\=&\nabla^\delta_{x,z}\cdot F^1_{\eta,v}+F^2_{\eta,v}.\end{aligned}$$

By Lemma 2.17 and Lemma 2.18, we have

$$\|T_{\nabla^\delta_{x,z}\widetilde{v}^h}\cdot\nabla\rho_\delta\|_{L^\infty_z(I;B^0_{\infty,1})}\le C\|\nabla_{x,z}\widetilde{v}\|_{L^\infty(\mathcal{S})}\|\nabla\rho_\delta\|_{L^\infty_z(I;B^0_{\infty,1})},$$
$$\|T_{\nabla\rho_\delta}\nabla^\delta_{x,z}\widetilde{v}^h\|_{L^\infty_z(I;B^0_{\infty,1})}\le C\|\nabla\rho_\delta\|_{L^\infty(\mathcal{S})}\|\nabla_{x,z}\widetilde{v}\|_{L^\infty_z(I;B^0_{\infty,1})},$$
$$\|R(\nabla^\delta_{x,z}\widetilde{v}^h,\nabla\rho_\delta)\|_{L^\infty_z(I;B^0_{\infty,1})}\le C\|\nabla_{x,z}\widetilde{v}\|_{L^\infty(\mathcal{S})}\|\nabla\rho_\delta\|_{L^\infty_z(I;B^\epsilon_{\infty,1})},$$

for any $\epsilon > 0$, which gives

$$\|\nabla_{x,z}^\delta \widetilde{v}^h \cdot \rho_\delta\|_{L_z^\infty(I;B_{\infty,1}^0)} \le K_\eta \|\widetilde{v}\|_{L_z^\infty(I;B_{\infty,1}^1)}.$$

The same estimate holds for $(\widetilde{v}^h, 0) f_\eta$. Hence,

$$\|F_{\eta,v}^1\|_{L_z^\infty(I;B_{\infty,1}^0)} \le K_\eta \|\widetilde{v}\|_{L_z^\infty(I;B_{\infty,1}^1)}.$$

By Lemma 2.17 and Lemma 2.18 again, we have

$$\begin{aligned}
&\|T_{\nabla_{x,z}^\delta \widetilde{v}^h} \nabla_{x,z}^\delta \nabla \rho_\delta\|_{\widetilde{L}_z^2(I;B_{\infty,1}^{-\frac{1}{2}})} \le C\|\nabla_{x,z}\widetilde{v}\|_{L^\infty(\overline{\mathcal{S}})} \|\nabla_{x,z}^\delta \rho_\delta\|_{\widetilde{L}_z^2(I;B_{\infty,1}^{\frac{1}{2}})},\\
&\|T_{\nabla_{x,z}^\delta \nabla \rho_\delta} \nabla_{x,z}^\delta \widetilde{v}^h\|_{\widetilde{L}_z^2(I;B_{\infty,1}^{-\frac{1}{2}})} \le C\|\nabla_{x,z}\widetilde{v}\|_{L_z^\infty(I;B_{\infty,1}^0)} \|\nabla_{x,z}^\delta \nabla \rho_\delta\|_{\widetilde{L}_z^2(I;B_{\infty,1}^{-\frac{1}{2}})},\\
&\|R(\nabla_{x,z}^\delta \nabla \rho_\delta, \nabla_{x,z}^\delta \widetilde{v}^h)\|_{\widetilde{L}_z^2(I;B_{\infty,1}^{-\frac{1}{2}})} \le C\|\nabla_{x,z}\widetilde{v}\|_{L^\infty(\overline{\mathcal{S}})} \|\nabla_{x,z}^\delta \rho_\delta\|_{\widetilde{L}_z^2(I;C^{1+\epsilon})},
\end{aligned}$$

for any $\epsilon > 0$, which gives

$$\|\nabla_{x,z}^\delta \widetilde{v}^h \nabla_{x,z}^\delta \nabla \rho_\delta\|_{\widetilde{L}_z^2(I;B_{\infty,1}^{-\frac{1}{2}})} \le K_\eta \|\nabla_{x,z}\widetilde{v}\|_{L^\infty(\overline{\mathcal{S}})}.$$

The same estimate holds for $\nabla \cdot \widetilde{v}^h f_\eta$ and $\partial_t f_\eta$. Hence,

$$\|F_{\eta,v}^2\|_{\widetilde{L}_z^2(I;B_{\infty,1}^{-\frac{1}{2}})} \le K_\eta \|\widetilde{v}\|_{W^{1,\infty}(\overline{\mathcal{S}})}.$$

Then we apply Proposition 4.2 to conclude that

$$\|\phi\|_{L_z^\infty(I;B_{\infty,1}^1)} \le K_\eta \|\widetilde{v}\|_{L_z^\infty(I;B_{\infty,1}^1)}. \tag{9.8}$$

Next we turn to the estimate of $\overline{v}^{d+1}$, which satisfies

$$\begin{aligned}
\Delta_{x,z}^\delta \overline{v}^{d+1} =& \Big(\frac{2|\nabla_{x,z}^\delta \partial_z \rho_\delta|^2}{(\partial_z \rho_\delta)^3} - \frac{\partial_z f_\eta}{(\partial_z \rho_\delta)^2}\Big)\phi\\
&+ 2\nabla_{x,z}^\delta \Big(\frac{1}{\partial_z \rho_\delta}\Big) \cdot \nabla_{x,z}\phi + \frac{1}{\partial_z \rho_\delta} \Delta_{x,z}^\delta \phi \triangleq G_{\eta,v}
\end{aligned}$$

with $\overline{v}_{d+1}|_{z=0} = 0$ and $\overline{v}_{d+1}|_{z=-1} = 0$. By Lemma 2.13 and Lemma 4.3, we have

$$\|G_{\eta,v}\|_{H^{s-\frac{3}{2}}(\overline{\mathcal{S}})} \le K_\eta \big(\|\phi\|_{H^{s+\frac{1}{2}}(\overline{\mathcal{S}})} + \|\nabla_{x,z}\phi\|_{L^\infty(\overline{\mathcal{S}})} \|\eta\|_{H^s}\big),$$

from which, (9.7)-(9.8) and Proposition 4.1, we deduce the first inequality.

We write

$$\begin{aligned}
G_{\eta,v} =& \Big(\frac{2|\nabla_{x,z}^\delta \partial_z \rho_\delta|^2}{(\partial_z \rho_\delta)^3} - \frac{\partial_z f_\eta}{(\partial_z \rho_\delta)^2}\Big)\phi + \nabla_{x,z}^\delta \Big(\frac{1}{\partial_z \rho_\delta}\Big) \cdot \nabla_{x,z}\phi + \nabla_{x,z}^\delta \cdot \Big(\frac{1}{\partial_z \rho_\delta} \nabla_{x,z}^\delta \phi\Big)\\
\triangleq& G_{\eta,v}^1 + \nabla_{x,z}^\delta \cdot G_{\eta,v}^2.
\end{aligned}$$

Then a similar argument leading to (9.8) yields

$$\|G_{\eta,v}^1\|_{\widetilde{L}_z^2(I;B_{\infty,1}^{-\frac{1}{2}})} + \|G_{\eta,v}^2\|_{L_z^\infty(I;B_{\infty,1}^0)} \le K_\eta \|\nabla_{x,z}\phi\|_{L_z^\infty(I;B_{\infty,1}^0)}.$$

Then Proposition 4.2 together with (9.8) gives the second inequality. $\square$

Now we are in position to prove Proposition 9.5.

PROOF OF PROPOSITION 9.5: Let $k \in [0, s-\frac{1}{2}]$ be an interger. Then we have

$$\frac{d}{dt}\|\nabla^k_{x,z}\widetilde{\omega}\|^2_{L^2(\mathcal{S})} = 2\big\langle -\nabla^k_{x,z}(\overline{v}\cdot\nabla_{x,z}\widetilde{\omega}) + \nabla^k_{x,z}F, \nabla^k_{x,z}\widetilde{\omega}\big\rangle.$$

First of all, we have by Lemma 2.13 that

$$\begin{aligned}\|\nabla^k_{x,z}F\|_{L^2(\mathcal{S})} \leq & K_\eta\|\nabla_{x,z}\widetilde{v}\|_{L^\infty(\mathcal{S})}\big(\|\widetilde{\omega}\|_{H^{s-\frac{1}{2}}(\mathcal{S})} + \|\nabla_{x,z}\widetilde{v}\|_{H^{s-\frac{1}{2}}(\mathcal{S})}\big)\\ & + K_\eta\|\nabla_{x,z}\widetilde{v}\|^2_{L^\infty(\mathcal{S})}\|\eta\|_{H^s}.\end{aligned}$$

Thanks to $\overline{v}^3 = 0$ on $z = 0$ and $z = -1$, we have

$$\big\langle\nabla^k_{x,z}(\overline{v}\cdot\nabla_{x,z}\widetilde{\omega}), \nabla^k_{x,z}\widetilde{\omega}\big\rangle = \big\langle[\nabla^k_{x,z}, \overline{v}]\cdot\nabla_{x,z}\widetilde{\omega}, \nabla^k_{x,z}\widetilde{\omega}\big\rangle - \big\langle \operatorname{div}_{x,z}\overline{v}\nabla^k_{x,z}\widetilde{\omega}, \nabla^k_{x,z}\widetilde{\omega}\big\rangle.$$

Then we deduece from Lemma 2.20 and Lemma 9.6 that

$$\begin{aligned}\big\langle\nabla^k_{x,z}(\overline{v}\cdot\nabla_{x,z}\widetilde{\omega}), \nabla^k_{x,z}\widetilde{\omega}\big\rangle \leq & C\|\nabla_{x,z}\overline{v}\|_{L^\infty(\mathcal{S})}\|\widetilde{\omega}\|_{H^{s-\frac{1}{2}}(\mathcal{S})}\|\nabla_{x,z}\overline{v}\|_{H^{s-\frac{1}{2}}(\mathcal{S})}\\ & + C\|\nabla_{x,z}\overline{v}\|_{L^\infty(\mathcal{S})}\|\widetilde{\omega}\|^2_{H^{s-\frac{1}{2}}(\mathcal{S})}\\ \leq & K_\eta A(t)\mathcal{B}(t)\big(\|\widetilde{\omega}\|^2_{H^{s-\frac{1}{2}}(\mathcal{S})} + \|\eta\|^2_{H^s} + \|\widetilde{v}\|^2_{H^{s+\frac{1}{2}}(\mathcal{S})}\big).\end{aligned}$$

Summing up, we conclude the proposition. □

9.4. Nonlinear estimates

Recall that the nonlinear term f is given by

$$f = D_t h_1 + (T_{a\lambda} - T_a T_\lambda)U + [T_a, D_t]\zeta + D_t[D_t, T_\zeta]B - T_a h_2,$$

where

$$\begin{aligned} h_1 &= (T_V - V)\cdot\nabla V - R(a,\zeta) + T_\zeta(T_V - V)\cdot\nabla B,\\ h_2 &= (T_V - V)\cdot\nabla\zeta + [T_\zeta, T_\lambda]B + (\zeta - T_\zeta)T_\lambda B + R(\eta)V + \zeta R(\eta)B.\end{aligned}$$

LEMMA 9.7. *It holds that*

$$\|h_1\|_{H^{s-\frac{1}{2}}} \leq K_\eta A(t)^2\big(\|V\|_{H^{s-\frac{1}{2}}} + \|\eta\|_{H^s}\big).$$

PROOF. It follows from Lemma 2.10 that

$$\|h_1\|_{H^{s-\frac{1}{2}}} \leq C\|\nabla V\|_{L^\infty}\|V\|_{H^{s-\frac{1}{2}}} + C\|a\|_{C^{\frac{1}{2}}}\|\zeta\|_{H^{s-1}} + K_\eta\|\nabla B\|_{L^\infty}\|V\|_{H^{s-\frac{1}{2}}},$$

which along with Lemma 7.8 gives the lemma. □

LEMMA 9.8. *It holds that*

$$\|D_t h_1\|_{H^{s-\frac{1}{2}}} \leq K_\eta\mathcal{B}(t)A(t)^2\big(\|(V,B)\|_{H^s} + \|\nabla_{x,z}\widetilde{v}\|_{L^2_z(I;H^{s-\frac{1}{2}})} + \|\eta\|_{H^{s+\frac{1}{2}}}\big).$$

PROOF. First, we consider the second term of the h_1. We denote $\overline{\partial}_t = \partial_t + V\cdot\nabla$. By Lemma 2.15, we get

$$\begin{aligned}\|\overline{\partial}_t R(a-1,\zeta)\|_{H^{s-\frac{1}{2}}} \leq & C\big(\|\overline{\partial}_t a\|_{L^\infty}\|\eta\|_{H^{s+\frac{1}{2}}} + \|\overline{\partial}_t\zeta\|_{L^\infty}\|a-1\|_{H^{s-\frac{1}{2}}}\\ & + \|\nabla V\|_{L^\infty}\big(\|a\|_{L^\infty}\|\eta\|_{H^{s+\frac{1}{2}}} + \|\zeta\|_{L^\infty}\|a-1\|_{H^{s-\frac{1}{2}}}\big) + \|V\|_{H^s}\|a\|_{L^\infty}\|\zeta\|_{C^{\frac{1}{2}+\epsilon}}\big).\end{aligned}$$

Using the fact that

$$\overline{\partial}_t a = D_t a + (V - T_V)\cdot\nabla a,$$

we get by Lemma 2.10 and Lemma 7.8 that

$$\|\overline{\partial}_t a\|_{L^\infty} \leq \|D_t a\|_{L^\infty} + \|V\|_{W^{1,\infty}}\|a\|_{L^\infty} \leq K_\eta\mathcal{B}(t)A(t)^2.$$

Note that $\overline{\partial}_t\zeta = \nabla B - \nabla V\cdot\nabla\eta$, hence,

$$\|\overline{\partial}_t\zeta\|_{L^\infty} \le K_\eta A(t).$$

Then by Lemma 7.8, we obtain

$$\|\overline{\partial}_t R(a-1,\zeta)\|_{H^{s-\frac12}} \le K_\eta \mathcal{B}(t)A(t)^2\big(\|\nabla_{x,z}\widetilde{v}\|_{L^2_z(I;H^{s-\frac12})} + \|V\|_{H^s} + \|\eta\|_{H^{s+\frac12}}\big).$$

which implies that

$$\begin{aligned}\|D_t R(a-1,\zeta)\|_{H^{s-\frac12}} \le& \|\overline{\partial}_t R(a-1,\zeta)\|_{H^{s-\frac12}} + C\|R(a-1,\zeta)\|_{C^{\frac12}}\|V\|_{H^s}\\ \le& K_\eta\mathcal{B}(t)A(t)^2\big(\|\nabla_{x,z}\widetilde{v}\|_{L^2_z(I;H^{s-\frac12})} + \|V\|_{H^s} + \|\eta\|_{H^{s+\frac12}}\big).\end{aligned}$$

First, we consider the first and third terms of the h_1. We write

$$D_t(T_V - V)\cdot\nabla V = -[D_t, T_{\nabla V}]V - T_{\nabla V}D_tV - D_tR(\nabla V, V).$$

By Remark 2.22, we have

$$\|[D_t, T_{\nabla V}]V\|_{H^{s-\frac12}} \le C\|V\|_{B^1_{\infty,1}}\|V\|_{H^{s-\frac12}} + C\|D_t\nabla V\|_{C^{-\frac12}}\|V\|_{H^s}.$$

While, using the equation

$$D_t\nabla V = \nabla D_tV - T_{\nabla V}\cdot\nabla V = \nabla(a\zeta) + \nabla(T_V - V)\cdot\nabla V - T_{\nabla V}\cdot\nabla V,$$

we deduce from Lemma 2.10 and Lemma 7.8 that

$$\|D_t\nabla V\|_{C^{-\frac12}} \le K_\eta\|a\|_{C^{\frac12}} + C\|V\|^2_{W^{1,\infty}} \le K_\eta A(t)^2.$$

This shows that

$$\|[D_t, T_{\nabla V}]V\|_{H^{s-\frac12}} \le C\|V\|_{B^1_{\infty,1}}\|V\|_{H^{s-\frac12}} + K_\eta A(t)^2\|V\|_{H^s}.$$

We get by Lemma 2.10 that

$$\|T_{\nabla V}D_tV\|_{H^{s-\frac12}} \le C\|V\|_{W^{1,\infty}}\|D_tV\|_{H^{s-\frac12}}.$$

It follows from Lemma 2.15 with $u = V$ and $v = \nabla V$ that

$$\begin{aligned}\|\overline{\partial}_t R(\nabla V, V)\|_{H^{s-\frac12}} \le& C\big(\|\overline{\partial}_t\nabla V\|_{C^{-\frac12}} + \|\overline{\partial}_tV\|_{C^{\frac12}} + \|V\|^2_{W^{1,\infty}}\big)\|V\|_{H^s}\\ \le& K_\eta A(t)^2\|V\|_{H^s},\end{aligned}$$

which ensures that

$$\begin{aligned}\|D_tR(\nabla V, V)\|_{H^{s-\frac12}} \le& \|\overline{\partial}_tR(\nabla V, V)\|_{H^{s-\frac12}} + \|R(\nabla V, V)\|_{C^{\frac12}}\|V\|_{H^s}\\ \le& K_\eta A(t)^2\|V\|_{H^s}.\end{aligned}$$

Thus, we obtain

$$\|D_t(T_V - V)\cdot\nabla V\|_{H^{s-\frac12}} \le K_\eta\mathcal{B}(t)A(t)^2\|V\|_{H^s}.$$

We write

$$D_tT_\zeta(T_V - V)\cdot\nabla B = [D_t, T_\zeta](T_V - V)\cdot\nabla B + T_\zeta D_t(T_V - V)\cdot\nabla B$$

In a similar way as the above, we can deduce that

$$\|D_tT_\zeta(T_V - V)\cdot\nabla B\|_{H^{s-\frac12}} \le K_\eta\mathcal{B}(t)A(t)^2\|(V, B)\|_{H^s}.$$

Putting the above estimates together gives the lemma. □

LEMMA 9.9. *It holds that*

$$\|h_2\|_{H^{s-\frac12}} \le K_\eta\mathcal{B}(t)A(t)\big(\|(V, B)\|_{H^s} + \|\eta\|_{H^{s+\frac12}}\big).$$

PROOF. By Lemma 2.10, we get

$$\|(T_V - V)\cdot\nabla\zeta\|_{H^{s-\frac12}} \le C\|\zeta\|_{C^{\frac12}}\|V\|_{H^s} \le K_\eta\|V\|_{H^s}.$$

By Proposition 2.6 and Lemma 4.6, we have

$$\|[T_\zeta, T_\lambda]B\|_{H^{s-\frac12}} \le K_\eta\|B\|_{H^s}.$$

It follows from Proposition 8.3 that

$$\|R(\eta)V\|_{H^{s-\frac12}} \le K_\eta\big(\|V\|_{H^s} + (\|\widetilde{v}\|_{L^\infty(I;B^1_{\infty,1})} + A(t))\|\eta\|_{H^{s+\frac12}}\big).$$

By Lemma 2.11 and Proposition 8.3, we get

$$\begin{aligned}\|\zeta R(\eta)B\|_{H^{s-\frac12}} \le& C\|R(\eta)B\|_{L^\infty}\|\eta\|_{H^{s+\frac12}} + C\|\zeta\|_{L^\infty}\|R(\eta)B\|_{H^{s-\frac12}}\\ \le& K_\eta\big(1 + \|\widetilde{v}\|_{L^\infty(I;B^1_{\infty,1})} + A(t)\big)\big(\|B\|_{H^s} + \|\eta\|_{H^{s+\frac12}}\big).\end{aligned}$$

By Lemma 2.10 and Proposition 2.7, we have

$$\|(\zeta - T_\zeta)T_\lambda B\|_{H^{s-\frac12}} \le C\|T_\lambda B\|_{L^\infty}\|\eta\|_{H^{s+\frac12}} \le K_\eta\|B\|_{B^1_{\infty,1}}\|\eta\|_{H^{s+\frac12}}.$$

This completes the proof of the lemma. □

LEMMA 9.10. *Let $f_1 = f - D_t[D_t, T_\zeta]B$. It holds that*

$$\|f_1\|_{H^{s-\frac12}} \le K_\eta\mathcal{B}(t)A(t)^3\big(\|(V,B)\|_{H^s} + \|\nabla_{x,z}\widetilde{v}\|_{L^2_z(I;H^{s-\frac12})} + \|\eta\|_{H^{s+\frac12}}\big).$$

PROOF. By Proposition 2.6, Lemma 7.8 and Lemma 4.6, we get

$$\begin{aligned}\|(T_{a\lambda} - T_aT_\lambda)U\|_{H^{s-\frac12}} \le& CM^1_{\frac12}(\lambda)M^0_{\frac12}(a)\|U\|_{H^s}\\ \le& K_\eta A(t)^2\|U\|_{H^s} \le K_\eta A(t)^2\|(V,B)\|_{H^s}.\end{aligned}$$

It follows from Proposition 2.21 and Lemma 7.8 that

$$\begin{aligned}\|[T_a, D_t]\zeta\|_{H^{s-\frac12}} \le& C\big(M^0_0(a)\|V\|_{B^1_{\infty,1}} + M^0_0(D_ta)\big)\|\eta\|_{H^{s+\frac12}}\\ \le& K_\eta\mathcal{B}(t)A(t)^2\|\eta\|_{H^{s+\frac12}}.\end{aligned}$$

By Lemma 2.10, Lemma 7.8 and Lemma 9.9, we get

$$\begin{aligned}\|T_ah_2\|_{H^{s-\frac12}} \le& C\|a\|_{L^\infty}\|h_2\|_{H^{s-\frac12}}\\ \le& K_\eta\mathcal{B}(t)A(t)^3\big(\|(V,B)\|_{H^s} + \|\eta\|_{H^{s+\frac12}}\big).\end{aligned}$$

Finally, the estimate for D_th_1 follows from Lemma 9.9. □

Next we estimate $f_\omega = -T_aR_\omega$, where R_ω is defined by

$$\begin{aligned}R^i_\omega =& \big(\partial_y v^i_\omega - \partial_{x_j}v^i_\omega\cdot\partial_{x_j}\eta\big) + \partial_{x_i}\eta\big(\partial_y v^{d+1}_\omega - \partial_{x_j}\eta\partial_{x_j}v^{d+1}_\omega\big)\\ &+ \big(\omega_{i,d+1} - \partial_{x_j}\eta\omega_{ij} + \partial_{x_i}\eta\partial_{x_j}\eta\omega_{j,d+1}\big)\big|_{y=\eta}.\end{aligned}$$

LEMMA 9.11. *It holds that*

$$\|f_\omega\|_{H^{s-1}} \le K_\eta A(t)^3\big(\|V_b\|_{H^{\frac12}} + \|\widetilde{\omega}\|_{H^{s-\frac12}(\mathcal{S})} + \|\eta\|_{H^s}\big).$$

PROOF. Let $I_0 = [-\frac12, 0]$. By Lemma 2.11 and Lemma 2.10, it is easy to see that

$$\begin{aligned}\|R_\omega\|_{H^{s-1}} \le& K_\eta\big(\|\nabla_{x,z}\widetilde{v_\omega}\|_{L^\infty(I_0;H^{s-1})} + \|\widetilde{\omega}\|_{L^\infty(I_0;H^{s-1})}\big)\\ &+ C\big(\|\nabla_{x,z}\widetilde{v_\omega}\|_{L^\infty_z(I_0;C^0)} + \|\omega\|_{L^\infty(\Omega_t)}\big)\|\eta\|_{H^s},\end{aligned}$$

from which and Lemma 8.4, we deduce that

$$\|R_\omega\|_{H^{s-1}} \le K_\eta A(t)\big(\|V_b\|_{H^{\frac12}} + \|\widetilde{\omega}\|_{H^{s-\frac12}(\mathcal{S})} + \|\eta\|_{H^s}\big). \tag{9.9}$$

Then the lemma follows from Lemma 2.10 and Lemma 7.8. □

LEMMA 9.12. *It holds that*

$$\|D_t f_\omega\|_{H^{s-1}} \le K_\eta \mathcal{B}(t) A(t)^4\big(\|(V,B,V_b)\|_{H^s} + \|\eta\|_{H^{s+\frac12}} + \|\widetilde{\omega}\|_{H^{s-\frac12}(\mathcal{S})}\big).$$

PROOF. Let $\overline{\partial}_t = \partial_t + V\cdot\nabla$. Then $\overline{\partial}_t R_\omega$ could be written as

$$\begin{aligned}\overline{\partial}_t R_\omega =& g_1(\nabla\eta)(\partial_t + v\cdot\nabla_{x,y})\nabla_{x,y}v_\omega + g_2(\nabla\eta)\overline{\partial}_t\nabla\eta\nabla_{x,z}v_\omega\\ &+ g_3(\nabla\eta)(\partial_t + v\cdot\nabla_{x,y})\omega + g_4(\nabla\eta)\overline{\partial}_t\nabla\eta\omega\big|_{y=\eta},\end{aligned}$$

for some smooth functions $g_i (i = 1,2,3,4)$. Then it follows from Lemma 2.11 that

$$\begin{aligned}\|\overline{\partial}_t R_\omega\|_{H^{s-1}} \le& K_\eta\big(\|(\partial_t + v\cdot\nabla_{x,y})\nabla_{x,y}v_\omega|_{y=\eta}\|_{H^{s-1}} + \|(\partial_t + v\cdot\nabla_{x,y})\omega|_{y=\eta}\|_{H^{s-1}}\big)\\ &+ K_\eta\big(\|(\partial_t + v\cdot\nabla_{x,y})\nabla_{x,y}v_\omega|_{y=\eta}\|_{C^0} + \|(\partial_t + v\cdot\nabla_{x,y})\omega\|_{L^\infty(\Omega_t)}\big)\|\eta\|_{H^{s+\frac12}}\\ &+ K_\eta\|\overline{\partial}_t\nabla\eta\|_{L^\infty}\big(\|\nabla_{x,y}v_\omega|_{y=\eta}\|_{H^{s-1}} + \|\omega|_{y=\eta}\|_{H^{s-1}}\big)\\ &+ K_\eta\big(\|\nabla_{x,y}v_\omega|_{y=\eta}\|_{L^\infty} + \|\omega\|_{L^\infty(\Omega_t)}\big)\|\overline{\partial}_t\nabla\eta\|_{H^{s-1}}\\ &+ K_\eta\big(\|\nabla_{x,y}v_\omega|_{y=\eta}\|_{L^\infty} + \|\omega\|_{L^\infty(\Omega_t)}\big)\|\overline{\partial}_t\nabla\eta\|_{L^\infty}\|\eta\|_{H^{s+\frac12}}.\end{aligned}$$

Let $I_0 = [-\frac12, 0]$. By Lemma 2.11 again, we get

$$\begin{aligned}\|(\partial_t + v\cdot\nabla_{x,y})\nabla_{x,y}v_\omega|_{y=\eta}\|_{H^{s-1}} \le& K_\eta\Big(\|\nabla_{x,z}\widetilde{\dot{v}_\omega}\|_{X^{s-1}(I_0)}\\ &+ \|\nabla_{x,z}\widetilde{v}_\omega\|_{L^\infty_z(I_0;L^\infty)}\|\nabla_{x,z}\widetilde{v}\|_{X^{s-1}(I_0)}\\ &+ \|\nabla_{x,z}\widetilde{v}_\omega\|_{X^{s-1}(I_0)}\|\nabla_{x,y}v\|_{L^\infty(\Omega_t)}\\ &+ \|\nabla_{x,z}\widetilde{v}_\omega\|_{L^\infty_z(I_0;L^\infty)}\|\nabla_{x,y}v\|_{L^\infty(\Omega_t)}\|\eta\|_{H^s}\Big).\end{aligned}$$

Here $\dot{v}_\omega = (\partial_t + v\cdot\nabla_{x,y})v_\omega$. Using the equation $(\partial_t + v\cdot\nabla_{x,y})\omega = \omega\cdot\nabla_{x,y}v$ and Lemma 2.11, we infer that

$$\begin{aligned}\|(\partial_t + v\cdot\nabla_{x,y})\omega|_{y=\eta}\|_{H^{s-\frac12}} \le& K_\eta\big(\|\nabla_{x,y}v\|_{L^\infty(\Omega_t)}\|\nabla_{x,z}\widetilde{v}\|_{X^{s-1}(I_0)}\\ &+ \|\nabla_{x,y}v\|^2_{L^\infty(\Omega_t)}\|\eta\|_{H^s}\big).\end{aligned}$$

Using the equation $\overline{\partial}_t\nabla\eta = \nabla B - \nabla V\cdot\nabla\eta$ and Lemma 2.11, we get

$$\|\overline{\partial}_t\nabla\eta\|_{H^{s-1}} \le K_\eta\big(\|(V,B)\|_{H^s} + \|\nabla V\|_{L^\infty}\|\eta\|_{H^s}\big).$$

On the other hand, we have

$$\begin{aligned}&\|(\partial_t + v\cdot\nabla_{x,y})\nabla_{x,y}v_\omega|_{y=\eta}\|_{C^0}\\ &\quad\le K_\eta\big(\|\nabla_{x,z}\widetilde{\dot{v}_\omega}\|_{L^\infty_z(I_0;C^0)} + \|\nabla_{x,z}\widetilde{v_\omega}\|_{L^\infty_z(I_0;L^\infty)}\|\nabla_{x,y}v\|_{L^\infty(\Omega_t)}\big),\\ &\|(\partial_t + v\cdot\nabla_{x,y})\omega|_{y=\eta}\|_{L^\infty} \le K_\eta\|\nabla_{x,y}v\|^2_{L^\infty(\Omega_t)},\\ &\|\overline{\partial}_t\nabla\eta|_{y=\eta}\|_{L^\infty} \le K_\eta\|(\nabla B, \nabla V)\|_{L^\infty}.\end{aligned}$$

Summing up the above estimates, we apply Lemma 8.4 and Lemma 8.5 to obtain

$$\|\overline{\partial}_t R_\omega\|_{H^{s-1}} \le K_\eta\mathcal{B}(t)A(t)^2\big(\|(V,B,V_b)\|_{H^s} + \|\eta\|_{H^{s+\frac12}} + \|\widetilde{\omega}\|_{H^{s-\frac12}(\mathcal{S})}\big),$$

which implies that

$$\begin{aligned}\|D_tR_\omega\|_{H^{s-1}}\leq&\|\overline{\partial}_tR_\omega\|_{H^{s-1}}+C\|R_\omega\|_{L^\infty}\|V\|_{H^s}\\ \leq&K_\eta\mathcal{B}(t)A(t)^2\big(\|(V,B,V_b)\|_{H^s}+\|\eta\|_{H^{s+\frac12}}+\|\widetilde{\omega}\|_{H^{s-\frac12}(\mathcal{S})}\big).\end{aligned}$$

which along with Proposition 2.21, Lemma 7.8 and (9.9) gives

$$\begin{aligned}\|D_tf_\omega\|_{H^{s-1}}\leq&\|[D_t,T_a]R_\omega\|_{H^{s-1}}+\|T_aD_tR_\omega\|_{H^{s-1}}\\ \leq&K_\eta\mathcal{B}(t)A(t)^4\big(\|(V,B,V_b)\|_{H^s}+\|\eta\|_{H^{s+\frac12}}+\|\widetilde{\omega}\|_{H^{s-\frac12}(\mathcal{S})}\big).\end{aligned}$$

The proof is finished. □

LEMMA 9.13. *It holds that*

$$\begin{aligned}\big\langle(f_\omega)_{s-1/2},(D_tU)_{s-1/2}\big\rangle\leq&\frac{d}{dt}\big\langle(f_\omega)_{s-1},U_s\big\rangle+K_\eta\mathcal{B}(t)A(t)^4\big(\|(V,B,V_b)\|_{H^s}\\ &+\|\eta\|_{H^{s+\frac12}}+\|\widetilde{\omega}\|_{H^{s-\frac12}(\mathcal{S})}\big)\|U\|_{H^s}.\end{aligned}$$

PROOF. A direct calculation yields

$$\begin{aligned}\big\langle(f_\omega)_{s-1/2},(D_tU)_{s-1/2}\big\rangle=&-\big\langle(f_\omega)_{s-1/2},[T_V\cdot\nabla,\langle D\rangle^{s-1/2}]U\big\rangle\\ &+\big\langle(f_\omega)_{s-1/2},D_tU_{s-1/2}\big\rangle\\ =&-\big\langle(f_\omega)_{s-1/2},[T_V\cdot\nabla,\langle D\rangle^{s-1/2}]U\big\rangle+\frac{d}{dt}\big\langle(f_\omega)_{s-1},U_s\big\rangle\\ &-\big\langle[T_V\cdot\nabla,\langle D\rangle^{s-1/2}]f_\omega,U_{s-1/2}\big\rangle-\big\langle(D_tf_\omega)_{s-1},U_s\big\rangle\\ &+\big\langle((T_V\cdot\nabla)^*+T_V\cdot\nabla)(f_\omega)_{s-1/2},U_{s-1/2}\big\rangle.\end{aligned}$$

By Lemma 2.19 and Proposition 2.6, we have

$$\begin{aligned}&\|[T_V\cdot\nabla,\langle D\rangle^{s-\frac12}]U\|_{H^{\frac12}}\leq C\|V\|_{W^{1,\infty}}\|U\|_{H^s},\\ &\|[T_V\cdot\nabla,\langle D\rangle^{s-\frac12}]f_\omega\|_{H^{-\frac12}}\leq C\|V\|_{W^{1,\infty}}\|f_\omega\|_{H^{s-1}},\\ &\|(T_V\cdot\nabla)^*+T_V\cdot\nabla)(f_\omega)_{s-1/2}\|_{H^{-\frac12}}\leq C\|V\|_{W^{1,\infty}}\|f_\omega\|_{H^{s-1}}.\end{aligned}$$

Then the lemma follows from Lemma 9.11 and Lemma 9.12. □

LEMMA 9.14. *It holds that*

$$\begin{aligned}&\big\langle(D_t[D_t,T_\zeta]B)_{s-\frac12},(D_tU)_{s-1/2}\big\rangle=\frac{d}{dt}\Big(\langle g_{s-1/2},h_{s-1/2}\rangle+\frac12\langle g_{s-1/2},g_{s-1/2}\rangle\Big)\\ &\quad+K_\eta\mathcal{B}(t)A(t)^3\big(\|D_tU\|_{H^{s-\frac12}}+\|(V,B)\|_{H^s}+\|\nabla_{x,z}\widetilde{v}\|_{L^2_z(I;H^{s-\frac12})}+\|\eta\|_{H^{s+\frac12}}\big)^2.\end{aligned}$$

where $g=[D_t,T_\zeta]B$ *and* $h=D_tU-[D_t,T_\zeta]B$.

PROOF. A direct calculation yields

$$\begin{aligned}
&\big\langle (D_t g)_{s-1/2}, (D_t U)_{s-1/2}\big\rangle = \big\langle (D_t g)_{s-1/2}, h_{s-1/2}\big\rangle + \big\langle (D_t g)_{s-1/2}, g_{s-1/2}\big\rangle\\
&= \frac{d}{dt}\Big(\big\langle g_{s-1/2}, h_{s-1/2}\big\rangle + \frac{1}{2}\big\langle g_{s-1/2}, g_{s-1/2}\big\rangle\Big)\\
&\quad - \big\langle [T_V\cdot\nabla, \langle D\rangle^{s-1/2}]g, h_{s-1/2}\big\rangle + \big\langle g_{s-1/2}, (D_t)^* h_{s-1/2}\big\rangle\\
&\quad - \big\langle [T_V\cdot\nabla, \langle D\rangle^{s-1/2}]g, g_{s-1/2}\big\rangle + \frac{1}{2}\big\langle ((T_V\cdot\nabla)^* + T_V\cdot\nabla) g_{s-1/2}, g_{s-1/2}\big\rangle\\
&\geq \frac{d}{dt}\Big(\big\langle g_{s-1/2}, h_{s-1/2}\big\rangle + \frac{1}{2}\big\langle g_{s-1/2}, g_{s-1/2}\big\rangle\Big)\\
&\quad - \|[T_V\cdot\nabla, \langle D\rangle^{s-1/2}]g\|_{L^2}\big(\|h\|_{H^{s-\frac12}} + \|g\|_{H^{s-\frac12}}\big)\\
&\quad - \|(T_V\cdot\nabla)^* + T_V\cdot\nabla) g_{s-1/2}\|_{L^2}\|g\|_{H^{s-\frac12}} - \|g\|_{H^s}\|(D_t)^* h_{s-1/2}\|_{H^{-\frac12}}.
\end{aligned}$$

It follows from Proposition 2.21 that

$$\begin{aligned}
\|g\|_{H^{s-\frac12}} &\leq C\big(\|V\|_{W^{1,\infty}} + \|D_t\zeta\|_{L^\infty}\big)\|B\|_{H^{s-\frac12}} \leq K_\eta A(t)\|B\|_{H^{s-\frac12}},\\
\|g\|_{H^s} &\leq C\big(\|V\|_{W^{1,\infty}} + \|D_t\zeta\|_{L^\infty}\big)\|B\|_{H^s} \leq K_\eta A(t)\|B\|_{H^s},
\end{aligned}$$

from which, Proposition 2.6 and Lemma 2.19, we infer

$$\begin{aligned}
&\|[T_V\cdot\nabla, \langle D\rangle^{s-1/2}]g\|_{L^2} \leq K_\eta A(t)^2\|B\|_{H^{s-\frac12}},\\
&\|((T_V\cdot\nabla)^* + T_V\cdot\nabla) g_{s-1/2}\|_{L^2} \leq K_\eta A(t)^2\|B\|_{H^{s-\frac12}},\\
&\|h\|_{H^{s-\frac12}} \leq \|D_t U\|_{H^{s-\frac12}} + K_\eta A(t)\|B\|_{H^{s-\frac12}}.
\end{aligned}$$

We have by Lemma 2.19 and Lemma 2.10 that

$$\begin{aligned}
\|(D_t)^* h_{s-1/2}\|_{H^{-\frac12}} &\leq \|D_t h\|_{H^{s-1}} + \|[T_V\cdot\nabla, \langle D\rangle^{s-\frac12}]h\|_{L^2} + \|T_{\nabla\cdot V} h_{s-1/2}\|_{L^2}\\
&\leq \|D_t h\|_{H^{s-1}} + C\|V\|_{W^{1,\infty}}\|h\|_{H^{s-\frac12}}\\
&\leq \|D_t h\|_{H^{s-1}} + K_\eta A(t)^2\big(\|D_t U\|_{H^{s-\frac12}} + \|B\|_{H^{s-\frac12}}\big).
\end{aligned}$$

While, by the equation (6.12), we find

$$D_t h = D_t h_1 - [D_t, T_a]\zeta - T_a D_t\zeta.$$

Then we get by Lemma 9.8, Proposition 2.21 and Lemma 7.8 that

$$\|D_t h\|_{H^{s-1}} \leq K_\eta \mathcal{B}(t)A(t)^2\big(\|(V,B)\|_{H^s} + \|\nabla_{x,z}\widetilde{v}\|_{L^2_z(I;H^{s-\frac12})} + \|\eta\|_{H^{s+\frac12}}\big).$$

This shows that

$$\begin{aligned}
\|(D_t)^* h_{s-1/2}\|_{H^{-\frac12}} \leq K_\eta \mathcal{B}(t)A(t)^2\big(&\|D_t U\|_{H^{s-\frac12}} + \|(V,B)\|_{H^s}\\
&+ \|\nabla_{x,z}\widetilde{v}\|_{L^2_z(I;H^{s-\frac12})} + \|\eta\|_{H^{s+\frac12}}\big).
\end{aligned}$$

Summing up the above estimates, we conclude the lemma. □

9.5. Energy functional

We introduce an energy functional $\mathcal{E}_s(t)$ defined by

$$\begin{aligned}
\mathcal{E}_s(t) \stackrel{\text{def}}{=} &\|v(t)\|_{H^1(\omega(t))} + \|\eta(t)\|_{H^s} + \|(V,B)(t)\|_{H^{s-\frac12}} + \|V_b(t)\|_{H^s}\\
&+ \|\widetilde{\omega}(t)\|_{H^{s-\frac12}(\overline{\mathcal{S}})} + \|T_{\sqrt{a\lambda}}U(t)\|_{H^{s-\frac12}} + \|D_t U(t)\|_{H^{s-\frac12}}.
\end{aligned} \tag{9.10}$$

PROPOSITION 9.15. *It holds that*

$$\|U\|_{H^s} + \|(V,B)\|_{H^s} + \|\eta\|_{H^{s+\frac{1}{2}}} + \|\widetilde{v}\|_{H^{s+\frac{1}{2}}(\mathcal{S})} \le K_\eta A(t)^3 \mathcal{E}_s(t).$$

PROOF. It follows from Lemma 8.1 that

$$\|\widetilde{v}\|_{H^{s+\frac{1}{2}}(\mathcal{S})} \le K_\eta A(t)\big(\|(V,B,V_b)\|_{H^s} + \|\widetilde{\omega}\|_{H^{s-\frac{1}{2}}(\mathcal{S})} + \|\eta\|_{H^{s+\frac{1}{2}}}\big).$$

This together with Lemma 9.16 and Lemma 9.17 yields the result. □

LEMMA 9.16. *It holds that*

$$\|\eta\|_{H^{s+\frac{1}{2}}} \le K_\eta A(t)^2 \mathcal{E}_s(t).$$

PROOF. Recall that $D_t U = -T_a\zeta + h_1 + [D_t, T_\zeta]B$. Due to $a \ge c_0$, we have

$$\begin{aligned}\zeta =& T_{a^{-1}}T_a\zeta + (T_{a^{-1}}T_a - 1)\zeta\\ =& T_{a^{-1}}\big(-D_tU + h_1 + [D_t, T_\zeta]B\big) + (T_{a^{-1}}T_a - 1)\zeta\end{aligned}$$

which along with Proposition 2.6, Lemma 9.7, Proposition 2.21 and Lemma 7.8 yields

$$\|\zeta\|_{H^{s-\frac{1}{2}}} \le K_\eta \|D_tU\|_{H^{s-\frac{1}{2}}} + K_\eta A(t)^2\big(\|\eta\|_{H^s} + \|(V,B)\|_{H^{s-\frac{1}{2}}}\big).$$

This gives the lemma by recalling $\zeta = \nabla\eta$. □

LEMMA 9.17. *It holds that*

$$\|U\|_{H^s} + \|(V,B)\|_{H^s} \le K_\eta A(t)^3 \mathcal{E}_s(t).$$

PROOF. Recall that $U = V + T_\zeta B$. Hence,

$$\|V\|_{H^s} \le \|U\|_{H^s} + \|T_\zeta B\|_{H^s} \le K_\eta\big(\|U\|_{H^s} + \|B\|_{H^s}\big).$$

So, it is sufficient to consider B. We have

$$\text{div}U = \text{div}V + T_\zeta\cdot\nabla B + T_{\text{div}\zeta}B.$$

On the other hand, we have

$$\begin{aligned}\text{div}V =& \sum_{i=1}^d \partial_i v^i + \partial_i\eta\partial_y v^i\big|_{y=\eta}\\ =& -\partial_y v^{d+1} + \nabla\eta\cdot\nabla v^{d+1}\big|_{y=\eta} + \partial_i\eta\omega_{d+1,i}\big|_{y=\eta}\\ =& -G(\eta)B - \partial_y v_\omega^{d+1} + \nabla\eta\cdot\nabla v_\omega^{d+1}\big|_{y=\eta} + \partial_i\eta\omega_{d+1,i}\big|_{y=\eta}\\ \triangleq& -G(\eta)B + V_\omega.\end{aligned}$$

Then we deduce that

$$\begin{aligned}\text{div}U =& \text{div}V + T_\zeta\cdot\nabla B + T_{\text{div}\zeta}B\\ =& -G(\eta)B + V_\omega + T_\zeta\cdot\nabla B + T_{\text{div}\zeta}B\\ =& -T_\lambda B - R(\eta)B + V_\omega + T_\zeta\cdot\nabla B + T_{\text{div}\zeta}B\\ =& -T_q B - R(\eta)B + V_\omega + T_{\text{div}\zeta}B,\end{aligned}$$

where the symbol $q = \lambda - i\zeta \cdot \xi$. Thus, it follows from Proposition 8.3, Lemma 8.4 and Lemma 9.16 that

$$\begin{aligned}\|T_q B\|_{H^{s-1}} \le& \|U\|_{H^s} + \|R(\eta)B\|_{H^{s-1}} + \|V_\omega\|_{H^{s-1}} + K_\eta \|B\|_{H^{s-1}}\\ \le& \|U\|_{H^s} + K_\eta A(t)\big(\|(B, V_b)\|_{H^{s-1}} + \|\omega\|_{H^{s-\frac12}(\mathcal{S})} + \|\eta\|_{H^{s+\frac12}}\big)\\ \le& \|U\|_{H^s} + K_\eta A(t)^3 \mathcal{E}(t).\end{aligned}$$

On the other hand, we get by Proposition 2.6 and Lemma 7.8 that

$$\begin{aligned}\|U\|_{H^s} \le& \|T_{(\sqrt{a\lambda})^{-1}} T_{\sqrt{a\lambda}} U\|_{H^s} + \|(T_{(\sqrt{a\lambda})^{-1}} T_{\sqrt{a\lambda}} - 1)U\|_{H^s}\\ \le& K_\eta A(t)^2 \|T_{\sqrt{a\lambda}} U\|_{H^{s-\frac12}}\| + K_\eta A(t)^2 \|U\|_{H^{s-\frac12}}\\ \le& K_\eta A(t)^2 \|T_{\sqrt{a\lambda}} U\|_{H^{s-\frac12}} + K_\eta A(t)^2 \|(V, B)\|_{H^{s-\frac12}},\end{aligned}$$

from which and Proposition 2.6, we infer that

$$\begin{aligned}\|B\|_{H^s} \le& K_\eta \|T_q B\|_{H^{s-1}} + \|(T_{q^{-1}} T_q - 1)B\|_{H^s}\\ \le& K_\eta A(t)^3 \mathcal{E}_s(t) + K_\eta \|B\|_{H^{s-\frac12}} \le K_\eta A(t)^3 \mathcal{E}_s(t).\end{aligned}$$

The proof is finished. □

9.6. Proof of Theorem 1.3

We first recover the regularity of the free surface from the mean curvature.

LEMMA 9.18. *Assume that the mean curvature $\kappa \in L^2 \cap L^p(\mathbf{R}^d)$ for some $p > d$. Then we have*

$$\|\eta\|_{H^2} + \|\eta\|_{C^{2-\frac{d}{p}}} \le C\big(\|\eta\|_{L^2}, \|\nabla\eta\|_{L^\infty}, \|H\|_{L^2\cap L^p}\big).$$

PROOF. The estimate of $\|\eta\|_{C^{2-\frac{d}{p}}}$ has been proved in [**33**]. Let $\eta_\ell = \partial_\ell \eta$ and $a_{ij} = (1 + |\nabla\eta|^2)^{-\frac32}\big((1 + |\nabla\eta|^2)\delta_{ij} - \partial_i\eta\partial_j\eta\big)$. A direct calculation gives

$$\partial_j\big(a_{ij}\partial_i\eta_\ell\big) = \partial_\ell H.$$

It is easy to verify that the matrix $\big(a_{ij}\big)$ is uniformly elliptic with the elliptic constants depending on $\|\nabla\eta\|_{L^\infty}$, which implies that $\|\nabla\eta_\ell\|_{L^2} \le C(\|\nabla\eta\|_{L^\infty})\|H\|_{L^2}$. □

LEMMA 9.19. *It holds that*

$$\sup_{t\in[0,T]} \big(A(t) + \|\eta(t)\|_{C^{\frac32+\varepsilon}}\big) \le C\big(T, M(T), \|v_0\|_{H^1(\Omega_0)}, \|\eta_0\|_{H^s}\big),$$

where C is an increasing function depending on h_0.

PROOF. Recall that $\zeta = \nabla\eta$ satisfies

$$\partial_t\zeta + V\cdot\nabla\zeta = \nabla B + \nabla V \cdot \zeta,$$

which implies

$$\sup_{t\in[0,T]} \|\nabla\eta(t)\|_{L^\infty} \le C\big(T, M(T), \|\nabla\eta_0\|_{L^\infty}\big),$$

which along with Lemma 9.1 and Lemma 9.18 implies that

$$\sup_{t\in[0,T]} \big(\|\eta(t)\|_{H^2} + \|\eta(t)\|_{C^{\frac32+\varepsilon}}\big) \le C\big(T, M(T), \|v_0\|_{L^2(\Omega_0)}, \|\eta_0\|_{H^s}\big). \tag{9.11}$$

By Lemma 9.1 and Lemma 9.2, we have

$$\frac{d}{dt}\big(\|v\|_{H^1(\Omega_t)}^2+\|\eta\|_{L^2}^2\big)\le K_\eta\big(\|v(t)\|_{W^{1,\infty}(\Omega_t)}\|v(t)\|_{H^1(\Omega_t)}+\|\eta(t)\|_{H^{\frac32}}\big),$$

from which and (9.11), we deduce the lemma. □

LEMMA 9.20. *Let $s>\frac{d}{2}+1$ and $f\in H^s(\mathbf{R}^d)$. Then there holds*

$$\|f\|_{B^1_{\infty,1}}\le C\big(1+\|f\|_{W^{1,\infty}}\big)\ln(e+\|f\|_{H^s}).$$

PROOF. Given an integer N, we get by Lemma 2.2 that

$$\begin{aligned}\|f\|_{B^1_{\infty,1}}=&\sum_{j\ge-1}^{N}2^j\|\Delta_jf\|_{L^\infty}+\sum_{j>N}2^j\|\Delta_jf\|_{L^\infty}\\ \le&C(N+1)\|f\|_{W^{1,\infty}}+C\sum_{j>N}2^{(1+\frac{d}{2})j}\|\Delta_jf\|_{L^2}\\ \le&C(N+1)\|f\|_{W^{1,\infty}}+C2^{-N(s-1-\frac{d}{2})}\|f\|_{H^s}.\end{aligned}$$

Take N so that $2^{-N(s-1-\frac{d}{2})}\|f\|_{H^s}\sim1$(i.e., $N\sim\ln(e+\|f\|_{H^s})$). Then the lemma follows easily. □

Now we are in position to prove Theorem 1.3. We denote

$$\mathcal{P}_T\triangleq\mathcal{P}\big(T,M(T),\|\nabla\eta_0\|_{L^\infty}\big)$$

for some increasing function $\mathcal{P}$ depending on c_0,h_0, which may change from line to line. By Lemma 9.19, K_η and $A(t)$ is bounded by $\mathcal{P}_T$. By Proposition 9.15, we have

$$\|U\|_{H^s}+\|(V,B)\|_{H^s}+\|\eta\|_{H^{s+\frac12}}+\|\widetilde{v}\|_{H^{s+\frac12}(\mathcal{S})}\le P_T\mathcal{E}_s(t).$$

We first deduce from Proposition 9.4 that

$$\frac{d}{dt}\big(\|D_tU\|^2_{H^{s-\frac12}}+\|T_{\sqrt{a\lambda}}U\|^2_{H^{s-\frac12}}\big)\le\mathcal{P}_T\mathcal{B}(t)\mathcal{E}_s(t)^2+2\langle(f+f_\omega)_{s-1/2},(D_tU)_{s-1/2}\rangle,$$

which along with Lemma 9.10, Lemma 9.13 and Lemma 9.14 gives

$$\begin{aligned}\frac{d}{dt}\Big(&\|D_tU\|^2_{H^{s-\frac12}}+\|T_{\sqrt{a\lambda}}U\|^2_{H^{s-\frac12}}+\big\langle(f_\omega)_{s-1},U_s\big\rangle\\ &+\big\langle g_{s-1/2},h_{s-1/2}\big\rangle+\frac12\big\langle g_{s-1/2},g_{s-1/2}\big\rangle\Big)\le\mathcal{P}_T\mathcal{B}(t)\mathcal{E}_s(t)^2,\end{aligned}$$

where $g=[D_t,T_\zeta]B$ and $h=D_tU-[D_t,T_\zeta]B$. By the proof of Lemma 9.14 and Lemma 9.11, we know

$$\begin{aligned}&\big\langle(f_\omega)_{s-1},U_s\big\rangle\le K_\eta A(t)^3\big(\|V_b\|_{H^{\frac12}}+\|\widetilde{\omega}\|_{H^{s-\frac12}}+\|\eta\|_{H^s}\big)\|U\|_{H^s},\\ &\big\langle g_{s-1/2},h_{s-1/2}\big\rangle\le K_\eta A(t)\|B\|_{H^{s-\frac12}}\|D_tU\|_{H^{s-\frac12}}+K_\eta A(t)^3\|B\|^2_{H^{s-\frac12}},\\ &\big\langle g_{s-1/2},g_{s-1/2}\big\rangle\le K_\eta A(t)^2\|B\|^2_{H^{s-\frac12}}.\end{aligned}$$

This shows that

$$\begin{aligned}\|D_tU(t)\|_{H^{s-\frac12}}+\|T_{\sqrt{a\lambda}}U(t)\|_{H^{s-\frac12}}\le&\,\mathcal{P}_T\mathcal{E}_s(0)+\mathcal{P}_0\int_0^t\mathcal{B}(t')\mathcal{E}_s(t')dt'\\ &+\mathcal{P}_T\big(\|(V,B)(t)\|_{H^{s-\frac12}}+\|V_b(t)\|_{H^s}+\|\widetilde{\omega}(t)\|_{H^{s-\frac12}(\mathcal{S})}+\|\eta(t)\|_{H^s}\big),\end{aligned}$$

which together with Proposition 9.5, Proposition 9.3 and Lemma 9.20 gives

$$\begin{aligned}\mathcal{E}_s(t) \leq & \mathcal{P}_0\mathcal{E}_s(0) + \mathcal{P}_T \int_0^t \mathcal{B}(t')\mathcal{E}_s(t')dt' \\ \leq & \mathcal{P}_T\mathcal{E}_s(0) + \mathcal{P}_T \int_0^t \ln(e + \mathcal{E}_s(t'))\mathcal{E}_s(t')dt'.\end{aligned}$$

Then Gronwall's inequality ensures that

$$\mathcal{E}_s(t) \leq C\big(E_s(0), \mathcal{P}_T\big).$$

This completes the proof of Theorem 1.3. □

CHAPTER 10

Iteration scheme

We start the proof of local well-posedness from this chapter. We first establish the local well-posedness result for sufficiently smooth data. In the last section, we will extend it to the low regularity data. Although there have been a lot of works [**27, 32, 40**] devoted to the local well-posedness of the Euler equations with free bondary for smooth data, these works do not work in the Eulerian coordinates and do not consider the case of finite depth to our knowledge.

THEOREM 10.1. *Let $d \geq 1$ and $s > \frac{d}{2} + 10$ be an integer. Assume that the initial data (η_0, v_0) satisfies*

$$\eta_0 \in H^{s+\frac{3}{2}}(\mathbf{R}^d), \quad v_0 \in H^{s+1}(\Omega_0).$$

Furthermore, we assume that there exist two positive constants $c_0 > 0$ and $h_0 > 0$ such that

$$\begin{aligned} &-(\partial_y P)(0, x, \eta_0(x)) \geq c_0 \quad \text{for } x \in \mathbf{R}^d, \\ &1 + \eta_0(x) \geq h_0 \quad \text{for } x \in \mathbf{R}^d. \end{aligned}$$

Then there exists $T > 0$ such that the system (1.1)–(1.5) with the initial data (η_0, v_0) has a unique solution (η, v) satisfying

$$\eta \in C\big([0,T]; H^{s+\frac{1}{2}}(\mathbf{R}^d)\big), \quad v \in C\big([0,T]; H^s(\Omega_t)\big).$$

REMARK 10.2. Here the initial data has more regularity than the solution. If we modify the first iteration, the solution can achieve the same regularity as the data.

The proof of Theorem 10.1 is conducted in the following four sections. In this section, we construct a sequence of approximate solutions by the iteration method. The next section is devoted to the uniform estimates of the approximate solutions. The following two sections are devoted to showing that the approximate sequence is a Cauchy sequence and converges to the solution of the Euler system (1.1)–(1.5).

The construction of approximate solutions is based on the following equivalent Euler system:

1. The evolution system on the free surface:

$$\begin{cases} D_t V = -T_\zeta a - T_a \zeta - R(\zeta, a) + (T_V - V) \cdot \nabla V, \\ D_t B = a - 1 + (T_V - V) \cdot \nabla B, \\ (\partial_t + V_b \cdot \nabla) V_b = -\nabla P|_{y=-1}, \\ D_t \zeta = T_\lambda (V + T_\zeta B) + (T_V - V) \cdot \zeta + [T_\zeta, T_\lambda] B \\ \qquad\qquad + (\zeta - T_\zeta) T_\lambda B + R(\eta) V + \zeta R(\eta) B + R_\omega, \\ (V, B, \zeta, V_b)\big|_{t=0} = \big(V_0, B_0, \nabla \eta_0, V_{b,0}\big), \end{cases} \tag{10.1}$$

where $D_t = \partial_t + T_V \cdot \nabla$, $\zeta = \nabla\eta$ and the i-component of R_ω is given by

$$
\begin{aligned}
(R_\omega)^i =& \big(\partial_y (v_\omega)^i - \partial_{x_j}(v_\omega)^i \cdot \partial_{x_j}\eta\big) + \partial_{x_i}\eta\big(\partial_y(v_\omega)^{d+1} - \partial_{x_j}\eta\partial_{x_j}(v_\omega)^{d+1}\big) \\
&+ \big(\omega_{i,d+1} - \partial_{x_j}\eta\omega_{ij} + \partial_{x_i}\eta\partial_{x_j}\eta\omega_{j,d+1}\big)\big|_{y=\eta},
\end{aligned}
$$

with v_ω defined by:

$$
\begin{cases} -\Delta_{x,y} v_\omega = \nabla_{x,y} \times \omega \quad \text{in} \quad \Omega_t, \\ v_\omega|_{y=\eta} = 0, \quad v_\omega|_{y=-1} = (V_b, 0). \end{cases}
$$

2. The evolution equation of vorticity:

$$
\begin{cases} \partial_t \omega + v \cdot \nabla\omega = \omega \cdot \nabla_{x,y} v, \\ \omega\,|_{t=0} = \omega_0, \end{cases} \tag{10.2}
$$

coupled with the equation of velocity v, which is defined by

$$
\begin{cases} -\Delta v = \nabla_{x,y} \times \omega \quad \text{in} \quad \Omega_t, \\ v\,|_{y=\eta} = (V, B), \\ v|_{y=-1} = (V_b, 0). \end{cases} \tag{10.3}
$$

3. The pressure P is determined by

$$
\begin{cases} -\Delta P = \partial_i v^j \partial_j v^i, \\ P\,|_{y=\eta} = 0, \quad \partial_y P\,|_{y=-1} = -1. \end{cases} \tag{10.4}
$$

10.1. Strategy

We use the following four steps to show the strategy of our construction.

Step 1. Iteration of the boundary values.

We denote by $\big(V^n, B^n, V_b^n, \eta^n, v^n, P^n, \omega^n\big)$ the n-th iteration. Then we define the $(n+1)$-th iteration. A natural idea is to introduce the following linear system:

$$
\begin{cases} (\partial_t + T_{V^n} \cdot \nabla)V^{n+1} + T_{a^n}\zeta^{n+1} = \Xi_1^n, \\ (\partial_t + T_{V^n} \cdot \nabla)B^{n+1} = \Xi_2^n, \\ (\partial_t + V_b^n \cdot \nabla)V_b^{n+1} = \Xi_3^n, \\ (\partial_t + T_{V^n} \cdot \nabla)\zeta^{n+1} - T_{\lambda^n}(V^{n+1} + T_{\zeta^n} B^{n+1}) = \Xi_4^n + R_\omega^n \\ \big(V^{n+1}, B^{n+1}, \zeta^{n+1}, V_b^{n+1}\big)\big|_{t=0} = \big(V_0, B_0, \nabla\eta_0, V_{b,0}\big), \end{cases}
$$

where Ξ_i^n is defined by $\big(V^n, B^n, V_b^n, \eta^n, v^n, P^n, \omega^n\big)$.

Once the approximate solutions were established, the next thing is to prove the uniform estimates of the approximate solutions. In this step, the second order system will be more useful, i.e.,

$$
\begin{aligned}
&(\partial_t + T_{V^n} \cdot \nabla)^2 \zeta^{n+1} + T_{\lambda^n} T_{a^n} \zeta^{n+1} \\
&= [(\partial_t + T_{V^n} \cdot \nabla), T_{\lambda^n}](V^{n+1} + T_{\zeta^n} B^{n+1}) + (\partial_t + T_{V^n} \cdot \nabla)(\Xi_4^n + R_\omega^n) \\
&\quad - T_{\lambda^n}(T_{\zeta^n}\Xi_2^n + \Xi_1^n - [(\partial_t + T_{V^n} \cdot \nabla), T_{\zeta^n}]B^{n+1})v + [T_{\lambda^n}, D_t](V^{n+1} + T_{\zeta^n} B^{n+1}).
\end{aligned} \tag{10.5}
$$

Let us remark that if we directly use the second order system (10.5) to construct the approximate solutions, it is not easy to recover (V, B) from the system. This is the reason why we choose the first order system.

To obtain the uniform estimates, we need the following two key facts:

- $(\partial_t + T_{V^n} \cdot \nabla)$ is a half order operator acting on Ξ_i^n, which means that if $\Xi_i^n \in H^{s-\frac{1}{2}}$, then $(\partial_t + T_{V^n} \cdot \nabla)\Xi_i^n \in H^{s-1}$;

- $(\partial_t + T_{V^n} \cdot \nabla) R^n_\omega$ has the same regularity as R^n_ω.

However, $(\partial_t + T_{V^n} \cdot \nabla)$ is an one order operator acting on Ξ^n_i. Thus, we have to modify the definition of Ξ^n_i. To this end, we introduce the auxiliary functions $(V^n_1, B^n_1, V^n_{b,1}, \eta^n_1)$ by solving the following linear transport system:

$$\begin{cases} (\partial_t + T_{V^n} \cdot \nabla)(V^n_1, B^n_1) = (\partial_t + T_{V^{n-1}} \cdot \nabla)(V^n, B^n), \\ (\partial_t + V^n_b \cdot \nabla) V^n_{b,1} = -\nabla P^{n-1}|_{y=-1}, \\ (V^n_1, B^n_1, V^n_{b,1})|_{t=0} = (V_0, B_0, V_{b,0}), \end{cases} \tag{10.6}$$

and η^n_1 is determined by

$$\begin{cases} (\partial_t + V^n \cdot \nabla)\eta^n_1 = B^n_1, \\ \eta^n_1|_{t=0} = \eta_0. \end{cases} \tag{10.7}$$

The key point is that although $(V^n_1, B^n_1, V^n_{b,1})$ loses $\frac{1}{2}$ regularity compared to (V^n, B^n, V^n_b), $(\partial_t + V^n \cdot \nabla)(V^n_1, B^n_1, V^n_{b,1})$ has the same regularity as $(V^n_1, B^n_1, V^n_{b,1})$. Moreover, $(\partial_t + V^n \cdot \nabla)^2 \eta^n_1, (\partial_t + V^n \cdot \nabla)\eta^n_1$ and η^n_1 have the same regularity. Then we redefine Ξ^n_i in terms of $(V^n_1, B^n_1, V^n_{b,1}, \eta^n_1)$ instead of $(V^n, B^n, V^n_b, \eta^n)$ in some places so that $(\partial_t + T_{V^n} \cdot \nabla)$ is a half order operator acting on Ξ^n_i.

Step 2. Iteration of the domain: η^{n+1}.

In the previous step, we define η^{n+1}_1, which can be used to determine the evolving domain Ω^{n+1}_t. However, compared to $\eta \in H^{s+\frac{1}{2}}$, η^{n+1}_1 lose one order regularity. However, we require that the domain belongs to $H^{s+\frac{1}{2}}$ in order to prove $\nabla P \in H^s(\Omega_t)$. To this end, we need to introduce a new domain, which has more regularity.

Now we give the definition of the $(n+1)$-th domain, which is defined by η^{n+1} from ζ^{n+1}. First of all, it is not easy to show that ζ^{n+1} is the gradient of some function. To avoid the loss of regularity, we define η^{n+1} by

$$(I - \Delta)\eta^{n+1} = -\mathrm{div}\,\zeta^{n+1} + \eta^{n+1}_1.$$

Thus, we introduce two fluid domains:

$$\Omega^{n+1}_t = \{(x, y) : -1 < y < \eta^{n+1}_1(t, x)\}$$

and

$$\widetilde{\Omega}^{n+1}_t = \{(x, y) : -1 < y < \eta^{n+1}(t, x)\}.$$

On one hand, $\widetilde{\Omega}^{n+1}_t$ is more regular than Ω^{n+1}_t. On the other hand, if we define v^{n+1}_2 with the boundary condition $v^{n+1}_2|_{\partial\Omega^{n+1}_t} = (V^{n+1}, B^{n+1}_1)$, then $\partial_t + v^{n+1}_2 \cdot \nabla_{x,y}$ is tangential to $\partial\Omega^{n+1}_t$. Two fluid domains have different good properties. This is the reason why we define the vorticity and the pressure in different fluid domains. Details will be discussed later.

Step 3. Iteration of the vorticity: ω^{n+1}.

Now we recover the vorticity by using (6.4) and (10.3). To construct ω^{n+1}, we need two important facts:

- $(\partial_t + V^{n+1} \cdot \nabla)(\omega^{n+1}|_{y=\eta^{n+1}_1})$ has the same regularity as $\omega^{n+1}|_{y=\eta^{n+1}_1}$;
- Material derivative is tangential to the boundary of domain.

The first fact is the key for the uniform estimates, while the second fact is necessary for the solvability of the equation of ω.

Now we define ω^{n+1} in the domain Ω_t^{n+1}:

$$\begin{cases} \partial_t \omega^{n+1} + ((v^{n+1})^h \cdot \nabla + (v_1^{n+1})^{d+1}\partial_y)\omega^{n+1} = \omega^{n+1} \cdot \nabla_{x,y} v_1^{n+1}, \\ \omega^{n+1}\,|_{t=0} = \omega_0, \end{cases} \tag{10.8}$$

where the velocity v^{n+1} is given by

$$\begin{cases} -\Delta v^{n+1} = \nabla_{x,y} \times \omega^{n+1} \quad \text{in} \quad \Omega_t^{n+1}, \\ v^{n+1}\,|_{y=\eta_1^{n+1}} = (V^{n+1}, B^{n+1}), \\ v^{n+1}|_{y=-1} = (V_b^{n+1}, 0), \end{cases} \tag{10.9}$$

and the velocity v_1^{n+1} is given by

$$\begin{cases} -\Delta v_1^{n+1} = \nabla_{x,y} \times \omega^{n+1} \quad \text{in} \quad \Omega_t^{n+1}, \\ v_1^{n+1}\,|_{y=\eta_1^{n+1}} = (V_1^{n+1}, B_1^{n+1}), \\ v_1^{n+1}|_{y=-1} = (V_{b,1}^{n+1}, 0). \end{cases} \tag{10.10}$$

This is a vorticity-velocity type system defined in a given moving domain. Although this is a nonlinear system, the boundary values and domain are given. We can prove the solvability of this system by similar arguments for the Euler equations in a fixed domain.

Let us explain why the above system meets our requirements. First of all, the boundary of ω^{n+1} is given by $y = \eta_1^{n+1}$, which satisfies

$$(\partial_t + V^n \cdot \nabla)\eta_1^n = B_1^n.$$

This implies that the material derivative $\partial_t + (v^{n+1})^h \cdot \nabla + (v_1^{n+1})^{d+1}\partial_y$ is tangential to the boundary of domain. By restricting the equation of ω^{n+1} to the boundary, we obtain

$$(\partial_t + V^{n+1} \cdot \nabla)(\omega^{n+1}|_{y=\eta_1^{n+1}}) = (\omega^{n+1} \cdot \nabla_{x,y} v_1^{n+1})|_{y=\eta_1^{n+1}},$$

which implies that $(\partial_t + V^{n+1} \cdot \nabla)(\omega^{n+1}|_{y=\eta_1^{n+1}})$ has the same regularity as $\omega^{n+1}|_{y=\eta_1^{n+1}}$.

On the right hand of (10.8), we use v_1^{n+1} instead of v^{n+1}. The reason is that although ω^{n+1} loses one half regularity compared to a priori estimate, ω^{n+1}, $(\partial_t + (v^{n+1})^h \cdot \nabla + (v_1^{n+1})^{d+1}\partial_y)\omega^{n+1}$ and $(\partial_t + (v^{n+1})^h \cdot \nabla + (v_1^{n+1})^{d+1}\partial_y)^2\omega^{n+1}$ have the same regularity.

Step 4. Iteration of the pressure: P^{n+1}.

To obtain $a^{n+1} \in H^{s-\frac{1}{2}}$, the fluid domain of the pressure at least belongs to $H^{s+\frac{1}{2}}$. Therefore, we need to choose $\widetilde{\Omega}_t^{n+1}$ as the fluid domain of the pressure. Since the velocity v_1^n is defined in the domain Ω_t^n, we introduce a map

$$\Phi_1^n : (x, z) \in \mathcal{S} \longmapsto (x, \rho_{\delta,1}^n(x, z)) \in \Omega_t^n,$$

with $\rho_{\delta,1}^n = z + (1+z)e^{\delta z|D|}\eta_1^n \in H^{s-1}$. Meanwhile,

$$\Phi^n : (x, z) \in \mathcal{S} \longmapsto (x, \rho_\delta^n(x, z)) \in \widetilde{\Omega}_t^n,$$

with $\rho_\delta^n = z + (1+z)e^{\delta z|D|}\eta^n$. Then $\big(\partial_i (v_1{}^j)^n \partial_j (v_1{}^i)^n\big) \circ \Phi_1^n \circ (\Phi^n)^{-1}$ is defined in $\widetilde{\Omega}_t^n$.

Now, we are in a position to determine P^{n+1}:

$$\begin{cases} -\Delta P^{n+1} = \left(\partial_i ({v_1}^j)^{n+1} \partial_j ({v_1}^i)^{n+1}\right) \circ \Phi_1^{n+1} \circ (\Phi^{n+1})^{-1} & \text{in } \widetilde{\Omega}_t^{n+1}, \\ P^{n+1}\mid_{y=\eta^{n+1}} = 0, \quad (\partial_y P^{n+1})\mid_{y=-1} = -1. \end{cases} \tag{10.11}$$

The following picture shows the relationship among the unknowns:

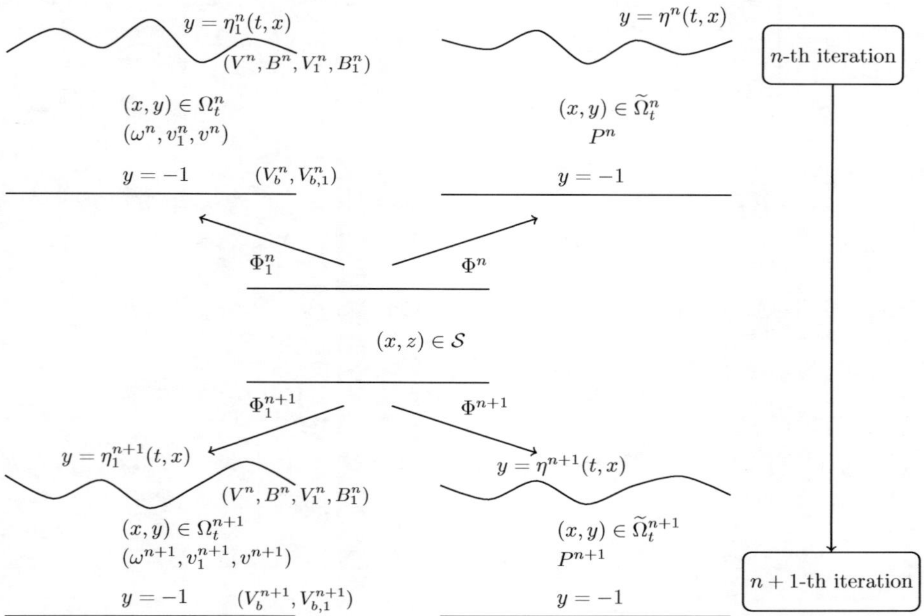

10.2. Iteration scheme

We start the iteration from initial data:

$$\left(V^0, B^0, V_b^0, \eta^0, v^0, P^0, \omega^0, V_1^0, B_1^0, V_{b,1}^0, \eta_1^0\right) = \left(V_0, B_0, V_{b,0}, \eta_0, v_0, P_0, \omega_0, V_0, B_0, V_{b,0}, \eta_0\right).$$

Assume that we have constructed a smooth solution $\left(V^n, B^n, V_b^n, \eta^n, P^n, \omega^n\right)$ and $\left(V_1^n, B_1^n, V_{b,1}^n, \eta_1^n\right)$ in n-th iteration in the time interval $[0, T_n]$ for $n \geq 0$, which satisfy $(H1) - (H8)$ in section 11.1. Here ω^n is a function defined in $\Omega_t^n = \{(x,y) : -1 < y < \eta_1^n(t,x)\}$ and P^n is a function defined in $\widetilde{\Omega}_t^n = \{(x,y) : -1 < y < \eta^n(t,x)\}$.

Let us construct the solution $\left(V^{n+1}, B^{n+1}, V_b^{n+1}, \eta^{n+1}, P^{n+1}, \omega^{n+1}\right)$ and $\left(V_1^{n+1}, B_1^{n+1}, V_{b,1}^{n+1}, \eta_1^{n+1}\right)$ in the $(n+1)$-th iteration by the following scheme.

We still denote by $D_t \triangleq \partial_t + T_{V^n} \cdot \nabla$ for the simplicity of notation. Let

$$a^n = -\partial_y P^n|_{y=\eta^n}, \quad \lambda^n = \lambda(\eta_1^n), \quad \zeta^n = \nabla \eta^n, \quad \zeta_1^n = \nabla \eta_1^n. \tag{10.12}$$

Step 1. Iteration of the boundary values.

We first introduce the evolution system on the trace of velocity and the free surface in the $(n+1)$-th iteration:

$$
\begin{cases}
D_t V^{n+1} + T_{a^n}\zeta^{n+1} = -T_{\zeta^n}a^n - R(\zeta^n, a^n) + (T_{V_1^n} - V_1^n)\cdot\nabla V^n,\\
D_t B^{n+1} = a^n - 1 + (T_{V_1^n} - V_1^n)\cdot\nabla B^n,\\
(\partial_t + V_b^n\cdot\nabla)V_b^{n+1} = -\nabla P^n|_{y=-1},\\
D_t\zeta^{n+1} - T_{\lambda^n}(V^{n+1} + T_{\zeta^n}B^{n+1}) = (T_{V_1^n} - V_1^n)\cdot\zeta^n + [T_{\zeta^n}, T_{\lambda^n}]B_1^n\\
\qquad + (\zeta^n - T_{\zeta^n})T_{\lambda^n}B^n + R(\eta_1^n)V_1^n + \zeta^n R(\eta_1^n)B_1^n + R_\omega^n,\\
(V^{n+1}, B^{n+1}, \zeta^{n+1}, V_b^{n+1})\big|_{t=0} = (V_0, B_0, \nabla\eta_0, V_{b,0}),
\end{cases}
\tag{10.13}
$$

where the i-component of R_ω^n is given by

$$
\begin{aligned}
(R_\omega^n)^i =&\big(\partial_y(v_\omega^n)^i - \partial_{x_j}(v_\omega^n)^i\cdot\partial_{x_j}\eta_1^n\big) + \partial_{x_i}\eta_1^n\big(\partial_y(v_\omega^n)^{d+1} - \partial_{x_j}\eta_1^n\partial_{x_j}(v_\omega^n)^{d+1}\big)\\
&+ \big(\omega_{i,d+1}^n - \partial_{x_j}\eta_1^n\omega_{ij}^n + \partial_{x_i}\eta_1^n\partial_{x_j}\eta_1^n\omega_{j,d+1}^n\big)\big|_{y=\eta_1^n}.
\end{aligned}
\tag{10.14}
$$

Here v_ω^n solves the following elliptic equation in ω_t^n:

$$
\begin{cases}
-\Delta_{x,y}v_\omega^n = \nabla_{x,y}\times\omega^n \quad \text{in} \quad \Omega_t^n,\\
v_\omega^n|_{y=\eta_1^n} = 0, \quad v_\omega^n|_{y=-1} = (V_{b,1}^n, 0).
\end{cases}
\tag{10.15}
$$

Given $(V^{n+1}, B^{n+1}, V_b^{n+1})$, we introduce a new boundary velocity $(V_1^{n+1}, B_1^{n+1}, V_{b,1}^{n+1})$ defined by

$$
\begin{cases}
(\partial_t + T_{V^{n+1}}\cdot\nabla)(V_1^{n+1}, B_1^{n+1}) = D_t(V^{n+1}, B^{n+1}),\\
(\partial_t + V_b^{n+1}\cdot\nabla)V_{b,1}^{n+1} = -\nabla P^n|_{y=-1},\\
(V_1^{n+1}, B_1^{n+1}, V_{b,1}^{n+1})\big|_{t=0} = (V_0, B_0, V_{b,0}).
\end{cases}
\tag{10.16}
$$

A key property is that (V_1^{n+1}, B_1^{n+1}) has the same regularity as $(\partial_t + T_{V^{n+1}}\cdot\nabla)(V_1^{n+1}, B_1^{n+1})$. While, $(\partial_t + T_{V^{n+1}}\cdot\nabla)$ is an one order operator acting on (V^{n+1}, B^{n+1}).

The system (10.13) can be symmetrized by introducing a good unknown $U^{n+1} = V^{n+1} + T_{\zeta^n}B^{n+1}$:

$$
\begin{cases}
D_t U^{n+1} + T_{a^n}\zeta^{n+1} = h_1^n + [D_t, T_{\zeta^n}]B^{n+1},\\
D_t\zeta^{n+1} = T_{\lambda^n}U^{n+1} + h_2^n + R_\omega^n,
\end{cases}
\tag{10.17}
$$

where (h_1^n, h_2^n) is given by

$$
\begin{aligned}
h_1^n =&(T_{V_1^n} - V_1^n)\cdot\nabla V^n - R(a^n, \zeta^n) + T_{\zeta^n}(T_{V_1^n} - V_1^n)\cdot\nabla B^n,\\
h_2^n =&(T_{V_1^n} - V_1^n)\cdot\nabla\zeta^n + [T_{\zeta^n}, T_{\lambda^n}]B_1^n + (\zeta^n - T_{\zeta^n})T_{\lambda^n}B^n\\
&+ R(\eta_1^n)V_1^n + \zeta^n R(\eta_1^n)B_1^n.
\end{aligned}
$$

A direct calculation shows that

$$
D_t^2\zeta^{n+1} + T_{a^n\lambda^n}\zeta^{n+1} = f_1^n + f_2^n + D_t f_3^n + D_t f_4^n + D_t f_\omega^n,
\tag{10.18}
$$

where $(f_1^n, f_2^n, f_3^n, f_4^n, f_\omega^n)$ are given by

$$
\begin{aligned}
f_1^n =& D_t\big((T_{V_1^n} - V_1^n)\cdot\nabla\zeta^n + [T_{\zeta^n}, T_{\lambda^n}]B_1^n\big) + (T_{a^n\lambda^n} - T_{\lambda^n}T_{a^n})\zeta^{n+1},\\
f_2^n =& [D_t, T_{\lambda^n}]U^{n+1} + T_{\lambda^n}\big(h_1^n + [D_t, T_{\zeta^n}]B^{n+1}\big),\\
f_3^n =& (\zeta^n - T_{\zeta^n})T_{\lambda^n}B^n + (\zeta^n - T_{\zeta^n})R(\eta_1^n)B_1^n,\\
f_4^n =& R(\eta_1^n)V_1^n + T_{\zeta^n}R(\eta_1^n)B_1^n,\\
f_\omega^n =& R_\omega^n.
\end{aligned}
$$

Step 2. Iteration of the domains: η^{n+1} and η_1^{n+1} .

Given $(V^{n+1}, B^{n+1}, \zeta^{n+1})$, we define η^{n+1} by

$$
-\Delta\eta^{n+1} + \eta^{n+1} = -\operatorname{div}\zeta^{n+1} + \eta_1^{n+1}, \tag{10.19}
$$

where η_1^{n+1} is determined by

$$
\begin{cases}
(\partial_t + V^{n+1}\cdot\nabla)\eta_1^{n+1} = B_1^{n+1},\\
\eta_1^{n+1}|_{t=0} = \eta_0.
\end{cases} \tag{10.20}
$$

This ends the first two steps. To make it clear, we give the following picture:

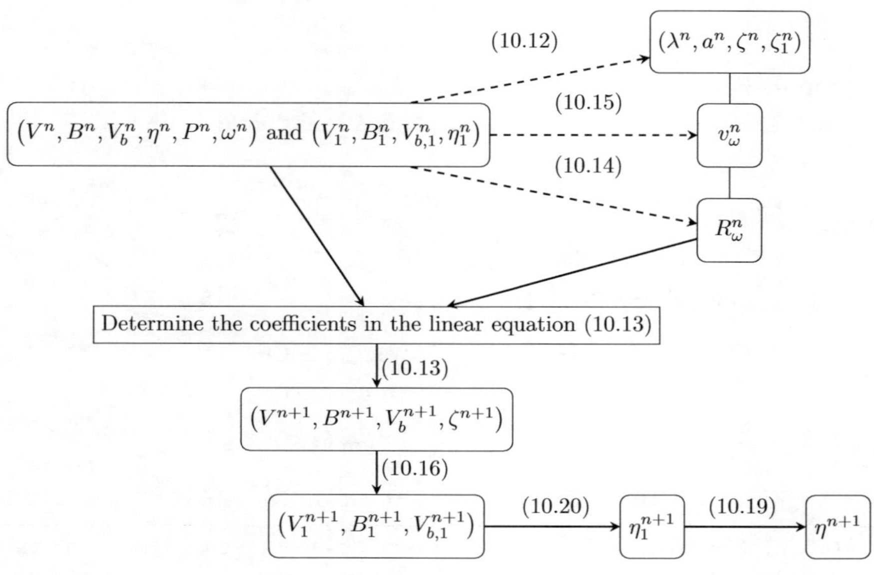

Step 3. Iteration of the vorticity: ω^{n+1}.

Given η_1^{n+1}, let $\Omega_t^{n+1} = \big\{(x,y) : -1 \le y \le \eta_1^{n+1}(t,x)\big\}$. The vorticity ω^{n+1} in $(n+1)$-th iterative is given by solving the nonlinear vorticity equation in Ω_t^{n+1}:

$$
\begin{cases}
\partial_t\omega^{n+1} + \big((v^{n+1})^h\cdot\nabla + (v_1^{n+1})^{d+1}\partial_y\big)\omega^{n+1} = \omega^{n+1}\cdot\nabla_{x,y}v_1^{n+1},\\
\omega^{n+1}\,|_{t=0} = \omega_0,
\end{cases} \tag{10.21}
$$

where the velocity v^{n+1} is given by

$$\begin{cases} -\Delta v^{n+1} = \nabla_{x,y} \times \omega^{n+1} \quad \text{in} \quad \Omega_t^{n+1}, \\ v^{n+1}\,|_{y=\eta_1^{n+1}} = (V^{n+1}, B^{n+1}), \\ v^{n+1}|_{y=-1} = (V_b^{n+1}, 0), \end{cases} \tag{10.22}$$

and the velocity v_1^{n+1} is given by

$$\begin{cases} -\Delta v_1^{n+1} = \nabla_{x,y} \times \omega^{n+1} \quad \text{in} \quad \Omega_t^{n+1}, \\ v_1^{n+1}\,|_{y=\eta_1^{n+1}} = (V_1^{n+1}, B_1^{n+1}), \\ v_1^{n+1}|_{y=-1} = (V_{b,1}^{n+1}, 0). \end{cases} \tag{10.23}$$

This ends the main steps. Here is a map which shows the relationship:

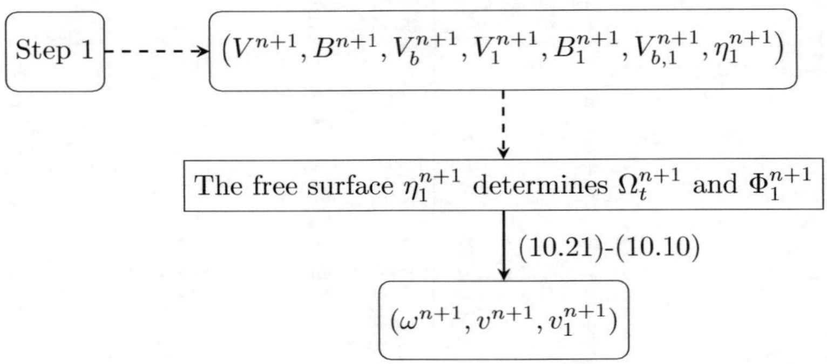

Step 4. Iteration of the pressure: P^{n+1}.

Finally, we construct the pressure P^{n+1} in a smoother domain $\widetilde{\Omega}_t^{n+1} = \{(x,y) : -1 < y < \eta^{n+1}(t,x)\}$:

$$\begin{cases} -\Delta P^{n+1} = \big(\partial_i ({v_1}^j)^{n+1} \partial_j ({v_1}^i)^{n+1}\big) \circ \Phi_1^{n+1} \circ (\Phi^{n+1})^{-1} \quad \text{in} \quad \widetilde{\Omega}_t^{n+1}, \\ P^{n+1}\,|_{y=\eta^{n+1}} = 0, \quad (\partial_y P^{n+1})\,|_{y=-1} = -1, \end{cases} \tag{10.24}$$

where the map Φ^{n+1} and Φ_1^{n+1} are given by

$$\Phi^{n+1} : (x,z) \in \mathcal{S} \longmapsto (x, \rho_\delta^{n+1}(x,z)) \in \widetilde{\Omega}_t^{n+1},$$
$$\Phi_1^{n+1} : (x,z) \in \mathcal{S} \longmapsto (x, \rho_{\delta,1}^{n+1}(x,z)) \in \Omega_t^{n+1}.$$

Here $\rho_\delta^{n+1} = \rho_{\delta,\eta^{n+1}}$ and $\rho_{\delta,1}^{n+1} = \rho_{\delta,\eta_1^{n+1}}$ with $\rho_{\delta,\eta}(x,z) = z + (1+z)e^{\delta z|D|}\eta$.

10.3. Existence of iteration scheme

10.3.1. The solvability of the linear system (10.13). We rewrite the approximate system of $(V^{n+1}, B^{n+1}, V_b^{n+1}, \zeta^{n+1})$ as follows

$$\begin{cases} (\partial_t + T_{V^n} \cdot \nabla) V^{n+1} + T_{a^n} \zeta^{n+1} = \Xi_1^n, \\ (\partial_t + T_{V^n} \cdot \nabla) B^{n+1} = \Xi_2^n, \\ (\partial_t + V_b^n \cdot \nabla) V_b^{n+1} = \Xi_3^n, \\ (\partial_t + T_{V^n} \cdot \nabla) \zeta^{n+1} - T_{\lambda^n}(V^{n+1} + T_{\zeta^n} B^{n+1}) = \Xi_4^n + R_\omega^n \\ \big(V^{n+1}, B^{n+1}, \zeta^{n+1}, V_b^{n+1}\big)\big|_{t=0} = \big(V_0, B_0, \nabla\eta_0, V_{b,0}\big). \end{cases}$$

Step 1. We define two Lagrangian flows:

$$\begin{cases} \partial_t X_b^n(t,x) = V_b^n(t, X_b^n(t,x))), & x \in \mathbf{R}^d, \\ X_b^n(0,x) = x, \end{cases}$$

and

$$\begin{cases} \partial_t X^n(t,x) = V^n(t, X^n(t,x))), & x \in \mathbf{R}^d, \\ X^n(0,x) = x. \end{cases}$$

Because V_b^n, V^n are given smooth functions, $X_b^n(t,x), X^n(t,x)$ exists on $[0,T_n]$. So, by the characteristic method, V_b^{n+1} exists in $[0,T_n]$. We rewrite the equation of B^{n+1} as

$$(\partial_t + V^n \cdot \nabla)B^{n+1} - F(V^n, B^{n+1}) = \Xi_2^n$$

where $F(V^n, B^{n+1}) = T_{\nabla B^{n+1}} V^n + R(\nabla B^{n+1}, V^n)$ satisfies

$$\|F\|_{H^s} \le C\|B^{n+1}\|_{H^s}.$$

So, B^{n+1} exists in $[0,T_n]$ by the characteristic method.

Step 2. The solvability of ζ^{n+1}, V^{n+1}.

Recall that

$$(\partial_t + T_{V^n}\cdot\nabla)\begin{pmatrix} V^{n+1} \\ \zeta^{n+1}\end{pmatrix} + \begin{pmatrix} 0 & T_{a^n} \\ -T_{\lambda^n} & 0 \end{pmatrix}\begin{pmatrix} V^{n+1} \\ \zeta^{n+1}\end{pmatrix} = \begin{pmatrix} \Xi_1^n \\ \Xi_4^n + R_\omega^n + T_{\lambda^n} T_{\zeta^n} B^{n+1}\end{pmatrix}.$$

We rewrite the above system as

$$\begin{aligned} &(\partial_t + V^n\cdot\nabla)\begin{pmatrix} V^{n+1} \\ \zeta^{n+1}\end{pmatrix} + \begin{pmatrix} 0 & T_{a^n} \\ -1-T_{\lambda^n} & 0 \end{pmatrix}\begin{pmatrix} V^{n+1} \\ \zeta^{n+1}\end{pmatrix} \\ &+\begin{pmatrix} (T_{V^n} - V^n)\cdot\nabla & 0 \\ 1 & (T_{V^n} - V^n)\cdot\nabla \end{pmatrix}\begin{pmatrix} V^{n+1} \\ \zeta^{n+1}\end{pmatrix} \\ &= \begin{pmatrix} \Xi_1^n \\ \Xi_4^n + R_\omega^n + T_{\lambda^n} T_{\zeta^n} B^{n+1}\end{pmatrix}. \end{aligned}$$

Acting $(1+T_{\lambda^n})^{-\frac12} T_{\sqrt{a^n}}$ on both sides of the equation of ζ^{n+1}, we obtain

$$\begin{aligned} &(\partial_t + V^n\cdot\nabla)\begin{pmatrix} V^{n+1} \\ (1+T_{\lambda^n})^{-\frac12} T_{\sqrt{a^n}}\zeta^{n+1}\end{pmatrix} \\ &+\begin{pmatrix} 0 & T_{\sqrt{a^n}}(1+T_{\lambda^n})^{\frac12} \\ -T_{\sqrt{a^n}}(1+T_{\lambda^n})^{\frac12} & 0 \end{pmatrix}\begin{pmatrix} V^{n+1} \\ (1+T_{\lambda^n})^{-\frac12} T_{\sqrt{a^n}}\zeta^{n+1}\end{pmatrix} \\ &+\mathfrak{R}(V^{n+1}, \zeta^{n+1}) = 0, \end{aligned}$$

where $\mathfrak{R}(V^{n+1}, \zeta^{n+1})$ satisfies

$$\|\mathfrak{R}(V^{n+1}, \zeta^{n+1})\|_{H^s} \le C(1 + \|V^{n+1}\|_{H^s} + \|\zeta^{n+1}\|_{H^s}).$$

Since the matrix

$$\begin{pmatrix} 0 & T_{\sqrt{a^n}}(1+T_{\lambda^n})^{\frac12} \\ -T_{\sqrt{a^n}}(1+T_{\lambda^n})^{\frac12} & 0 \end{pmatrix}$$

is a skew symmetric matrix, by the standard energy method, we obtain the solvability for V^{n+1} and ζ^{n+1} on $[0, T_n]$.

Step 3. Once we have $V^{n+1}, B^{n+1}, \zeta^{n+1}$ and V_b^{n+1} on $[0, T_n]$, by the characteristic method, we immediately get the solvability for $(V_1^{n+1}, B_1^{n+1}, \eta_1^{n+1}, V_{b,1}^{n+1})$ on $[0, T_n]$. It is easy to solve η^{n+1} because the operator $I - \Delta$ is reversible.

10.3.2. The solvability of the vorticity. The vorticity-velocity system is a nonlinear system in a domain depending on the time t. Here we present a sketch of the proof for the solvability.

To simplify the notation, we drop the indicator $n+1$. The fluid domain is given by

$$\Omega_t = \{y : -1 < y < \eta_1(t, x)\},$$

where η_1 is a given smooth function and satisfies that $1 + \eta_1 \geq \frac{h_0}{2}$ in $[0, T]$. The vorticity ω solves

$$\begin{cases} \partial_t \omega + \big((v)^h \cdot \nabla + (v_1)^{d+1} \partial_y\big)\omega = \omega \cdot \nabla_{x,y} v_1, \\ \omega \mid_{t=0} = \omega_0, \end{cases} \tag{10.25}$$

where the velocity v is given by

$$\begin{cases} -\Delta v = \nabla_{x,y} \times \omega \quad \text{in} \quad \Omega_t, \\ v \mid_{y=\eta_1} = (V, B), \\ v|_{y=-1} = (V_b, 0), \end{cases} \tag{10.26}$$

and the velocity v_1 is given by

$$\begin{cases} -\Delta v_1 = \nabla_{x,y} \times \omega \quad \text{in} \quad \Omega_t, \\ v_1 \mid_{y=\eta_1} = (V_1, B_1), \\ v_1|_{y=-1} = (V_{b,1}, 0). \end{cases} \tag{10.27}$$

Here, (V, B) and (V_1, B_1) are given in $[0, T]$ so that

$$\partial_t \eta_1 + V \cdot \nabla \eta_1 = B_1.$$

PROPOSITION 10.3. *There exists a positive time $T_1 \leq T$ such that the system* (10.25)–(10.27) *has a unique smooth solution (v_1, v, ω) in $[0, T_1]$.*

PROOF. The proof is split into the following steps.

Step 1. Formulation in a fixed domain .

We introduce a mapping Φ which is defined by

$$\Phi : (x, z) \in \mathcal{S} \longmapsto (x, \rho_{\delta,1}(x, z)) \in \Omega_t,$$

where $\rho_{\delta,1} = \rho_{\delta,\eta_1}$ with $\rho_{\delta,\eta_1}(x, z) = z + (1+z)e^{\delta z|D|}\eta_1$ and $\mathcal{S} = [-1, 0] \times \mathbf{R}^d$. We denote $\widetilde{f}(t, x, z) = f \circ \Phi$. In terms of the (x, z) variable, we have

$$\partial_t \widetilde{\omega} + \overline{v} \cdot \nabla_{x,z} \widetilde{\omega} = F, \tag{10.28}$$

where

$$\overline{v} = \Big((\widetilde{v})^h, \frac{1}{\partial_z \rho_{\delta,\eta_1}}\big(\widetilde{v_1}^{d+1} - \partial_t \rho_{\delta,\eta_1} - (\widetilde{v})^h \cdot \nabla \rho_{\delta,\eta_1}\big)\Big),$$

$$F = (\widetilde{\omega})^h \cdot \big(\nabla \widetilde{v_1} - \frac{\nabla \rho_{\delta,\eta_1}}{\partial_z \rho_{\delta,\eta_1}} \partial_z \widetilde{v_1}\big) + (\widetilde{\omega})^{d+1} \frac{\partial_z \widetilde{v_1}}{\partial_z \rho_{\delta,\eta_1}}.$$

Here one important property is that

$$\overline{v} \cdot e_z|_{z=0,-1} = 0.$$

Step 2. Approximate solutions.

We construct the approximate solution by an iteration. Assume that the initial data (ω_0, v_0) is smooth, and we construct a smooth solution $(\widetilde{v}^n, \widetilde{v}_1^n, \widetilde{\omega}^n)$ in n-th iteration. Next, we construct $(\widetilde{v}^{n+1}, \widetilde{v}_1^{n+1}, \widetilde{\omega}^{n+1})$ in the $(n+1)$-th iteration by the following scheme.

First of all, the solvability of v and v_1 is obvious. And $\widetilde{\omega}^{n+1}$ is defined by the following linear equation:

$$\partial_t \widetilde{\omega}^{n+1} + \overline{v}^n \cdot \nabla_{x,z} \widetilde{\omega}^{n+1} = F^n,$$

where

$$\overline{v}^n = \Big((\widetilde{v}^n)^h, \frac{1}{\partial_z \rho_{\delta,\eta_1}} ((\widetilde{v}_1{}^n)^{d+1} - \partial_t \rho_{\delta,\eta_1} - (\widetilde{v}^n)^h \cdot \nabla \rho_{\delta,\eta_1}) \Big),$$

$$F = (\widetilde{\omega}^n)^h \cdot \Big(\nabla \widetilde{v}_1{}^n - \frac{\nabla \rho_{\delta,\eta_1}}{\partial_z \rho_{\delta,\eta_1}} \partial_z \widetilde{v}_1{}^n \Big) + (\widetilde{\omega}^n)^{d+1} \frac{\partial_z \widetilde{v}_1{}^n}{\partial_z \rho_{\delta,\eta_1}}.$$

Because of $\overline{v}^n \cdot e_z|_{z=0,-1} = 0$, the above linear equation is solvable.

Step 3. Uniform estimates and Cauchy sequence.

This step is similar to the Euler equations in a fixed domain. □

CHAPTER 11

Uniform energy estimates

11.1. Set-up

Throughout this chapter, we denote

$$D_t \triangleq \partial_t + T_{V^n}\cdot\nabla, \quad D_t^u \triangleq \partial_t + V^n\cdot\nabla, \quad D_t^b \triangleq \partial_t + V_b^n\cdot\nabla, \quad \mathcal{D}_t \triangleq \partial_t + v_2^n\cdot\nabla_{x,y},$$

where $v_2^n = \big((v^n)^h, (v_1^n)^{d+1}\big)$. We denote

$$\widetilde{D}_t \triangleq \partial_t + T_{V^{n+1}}\cdot\nabla, \quad \widetilde{D}_t^u \triangleq \partial_t + V^{n+1}\cdot\nabla,$$
$$\widetilde{D}_t^b \triangleq \partial_t + V_b^{n+1}\cdot\nabla, \quad \widetilde{\mathcal{D}}_t \triangleq \partial_t + v_2^{n+1}\cdot\nabla_{x,y},$$

where $v_2^{n+1} = \big((v^{n+1})^h, (v_1^{n+1})^{d+1}\big)$. For a function $f(x,y)$ defined on $\big\{(x,y) : -1 < y < \eta(x)\big\}$, we denote $\widetilde{f}(x,z) \triangleq f(x,\rho_\delta(x,z))$, where $\rho_\delta(x,z) = z + (1+z)e^{\delta z|D|}\eta$.

Let $E_i (i = 1,\cdots,6)$ be some constants determined later. Assume that there exists T independent of n determined later such that the solution in the n-th iteration satisfies

H1. For any $t\in[0,T]$, there holds

$$\big\|(V^n, B^n, V_b^n, V_1^n, B_1^n, V_{b,1}^n)(t)\big\|_{H^{s-\frac12}} + \|(\eta^n,\eta_1^n)(t)\|_{H^{s-\frac12}} + \|\widetilde{\omega}^n(t)\|_{H^{s-1}(\mathcal{S})} \le E_1;$$

H2. For any $t\in[0,T]$, there holds

$$\|\nabla_{x,z}\widetilde{v_1^n}(t)\|_{H^{s-1}(\mathcal{S})} + \|\widetilde{v_\omega^n}(t)\|_{H^s(\mathcal{S})} + \|a^n(t)\|_{H^{s-\frac32}} + \|\nabla P^n(t,\cdot,-1)\|_{H^{s-\frac32}} \le E_2;$$

H3. For any $t\in[0,T]$, there holds

$$\|a^n(t)\|_{H^{s-\frac12}} + \|\nabla P^n(t,\cdot,-1)\|_{H^{s-\frac12}} \le E_3^1, \quad \|\eta^n(t)\|_{H^{s+\frac12}} \le E_3^2;$$

H4. For any $t\in[0,T]$, there holds

$$\begin{aligned}&\|D_t(V_1^n, B_1^n)(t)\|_{H^{s-\frac32}} + \|\partial_t\eta^n(t)\|_{H^{s-\frac32}} + \|D_t^u\eta_1^n(t)\|_{H^{s-\frac12}}\\ &+ \|\widetilde{\mathcal{D}_t\omega^n}(t)\|_{H^{s-1}(\mathcal{S})} + \|\widetilde{\mathcal{D}_t v_\omega^n}(t)\|_{X^{s-\frac12}([-\frac12,0])} \le E_4;\end{aligned}$$

H5. For any $t\in[0,T]$, there holds

$$\begin{aligned}&\|\big(D_t(V_1^n, B_1^n), D_t^b V_{b,1}^n\big)(t)\|_{H^{s-\frac12}} + \|\partial_t^2\eta^n(t)\|_{H^{s-\frac52}} + \|\partial_t a^n(t)\|_{H^{s-\frac32}}\\ &+ \|(D_t^u)^2\eta_1^n(t)\|_{H^{s-\frac12}} + \|\widetilde{\mathcal{D}_t^2\omega^n}(t)\|_{H^{s-1}(\mathcal{S})} + \|\nabla_{x,z}\widetilde{\mathcal{D}_t v_1^n}\|_{H^{s-1}(\mathcal{S})}\\ &+ \|\nabla\partial_t P^n(t,\cdot,-1)\|_{H^{s-\frac32}} \le E_5;\end{aligned}$$

H6. For any $t\in[0,T]$, there holds

$$\|\partial_t\big(D_t(V_1^n, B_1^n), D_t^b V_{b,1}^n\big)(t)\|_{H^{s-\frac32}} + \|\widetilde{\mathcal{D}_t^2 v_\omega^n}(t)\|_{X^{s-\frac32}([-\frac12,0])} \le E_6;$$

H7. For any $(t,x) \in [0,T] \times \mathbf{R}^d$, there holds

$$a^n(t,x) = -\partial_y P^n\big|_{y=\eta^n(t,x)} \geq \frac{c_0}{2};$$

H8. For any $(t,x) \in [0,T] \times \mathbf{R}^d$, there holds

$$\eta^n(t,x) + 1 \geq \frac{h_0}{2}, \quad \eta_1^n(t,x) + 1 \geq \frac{h_0}{2}.$$

The purpose of this section is to show that (H1)–(H8) also hold for the solution in the $(n+1)$-th iteration.

In the sequel, we denote $\mathcal{A}_k = \mathcal{A}_k\big(E_1, \cdots, E_k\big)$ for $k = 1, \cdots, 6$, and $P_{s,\eta_1,\cdots,\eta_k} = P_{s,\eta_1,\cdots,\eta_k}\big(\|\eta_1\|_{H^s}, \cdots, \|\eta_k\|_{H^s}\big)$, where $\mathcal{A}_k$ and $P_{s,\eta_1,\cdots,\eta_k}$ are some nondecreasing functions depending on c_0, h_0 and may change from line to line. We also denote by $\mathcal{P}(\cdot, \cdots, \cdot)$ some increasing function depending on c_0, h_0, which may be different from line to line.

11.2. Energy functional

We introduce the energy functional $\mathcal{E}^{n+1}(t)$ defined by

$$\mathcal{E}^{n+1}(t) \stackrel{\text{def}}{=} \mathcal{E}_1^{n+1}(t) + \mathcal{E}_2^{n+1}(t),$$

where $\mathcal{E}_1^{n+1}(t)$ and $\mathcal{E}_2^{n+1}(t)$ are given by

$$\begin{aligned}
\mathcal{E}_1^{n+1}(t) =& \|D_t\zeta^{n+1}\|_{H^{s-1}} + \|T_{\sqrt{a^n\lambda^n}}\zeta^{n+1}\|_{H^{s-1}},\\
\mathcal{E}_2^{n+1}(t) =& \|\widetilde{\omega}^{n+1}\|_{H^{s-1}(\overline{\mathcal{S}})} + \|\zeta^{n+1}\|_{H^{s-\frac{3}{2}}} + \|(V^{n+1}, B^{n+1}, V_b^{n+1})\|_{H^{s-\frac{1}{2}}}\\
&+ \|(V_1^{n+1}, B_1^{n+1}, V_{b,1}^{n+1}, \eta_1^{n+1})\|_{H^{s-\frac{1}{2}}}.
\end{aligned}$$

Using the equations, we can establish the following regularity information in terms of $\mathcal{E}^{n+1}(t)$ for the solution in the $(n+1)$-th iteration.

Lemma 11.1. *It holds that*

$$\|\eta^{n+1}\|_{H^{s+\frac{1}{2}}} \leq \mathcal{A}_2\mathcal{E}^{n+1}(t), \quad \|U^{n+1}\|_{H^{s-\frac{1}{2}}} \leq \mathcal{A}_1\mathcal{E}_2^{n+1}(t).$$

Proof. We write

$$\zeta^{n+1} = T_{(\sqrt{a^n\lambda^n})^{-1}}T_{\sqrt{a^n\lambda^n}}\zeta^{n+1} + \big(T_{(\sqrt{a^n\lambda^n})^{-1}}T_{\sqrt{a^n\lambda^n}} - 1\big)\zeta^{n+1},$$

from which and Proposition 2.6, we infer that

$$\begin{aligned}
\|\zeta^{n+1}\|_{H^{s-\frac{1}{2}}} \leq& \mathcal{A}_2\|\zeta^{n+1}\|_{H^{s-\frac{3}{2}}} + \mathcal{A}_2\|T_{\sqrt{a^n\lambda^n}}\zeta^{n+1}\|_{H^{s-1}}\\
\leq& \mathcal{A}_2\mathcal{E}^{n+1}(t).
\end{aligned}$$

Recall that η^{n+1} satisfies

$$-\Delta\eta^{n+1} + \eta^{n+1} = -\operatorname{div}\zeta^{n+1} + \eta_1^{n+1},$$

which implies that

$$\|\eta^{n+1}\|_{H^{s+\frac{1}{2}}} \leq \|\zeta^{n+1}\|_{H^{s-\frac{1}{2}}} + \|\eta_1^{n+1}\|_{H^{s-\frac{3}{2}}} \leq \mathcal{A}_2\mathcal{E}^{n+1}(t).$$

The second inequality follows from $U^{n+1} = V^{n+1} + T_{\zeta^n}B^{n+1}$. □

Lemma 11.2. *It holds that*

$$\|D_tU^{n+1}\|_{H^{s-\frac{1}{2}}} \leq \mathcal{A}_1 + \mathcal{A}_4\mathcal{E}^{n+1}(t).$$

PROOF. Recall that

$$D_tU^{n+1}+T_{a^n}\zeta^{n+1}=h_1^n+[D_t,T_{\zeta^n}]B^{n+1}. \tag{11.1}$$

By Lemma 2.10, we have

$$\|h_1^n\|_{H^{s-\frac12}}\le\mathcal{A}_1,$$

which along with Proposition 2.21 and Lemma 11.1 gives

$$\begin{aligned}\|D_tU^{n+1}\|_{H^{s-\frac12}}\le&\mathcal{A}_1+\mathcal{A}_2\|\zeta^{n+1}\|_{H^{s-\frac12}}+\mathcal{A}_4\|B^{n+1}\|_{H^{s-\frac12}}\\ \le&\mathcal{A}_1+\mathcal{A}_4\mathcal{E}^{n+1}(t).\end{aligned}$$

The proof is finished. □

LEMMA 11.3. *It holds that*

$$\begin{aligned}&\|\widetilde{D}_t^u\eta_1^{n+1}\|_{H^{s-\frac12}}\le\mathcal{E}_2^{n+1}(t),\\ &\|(\widetilde{D}_t^u)^2\eta_1^{n+1}\|_{H^{s-\frac12}}\le\mathcal{A}_3\big(1+\mathcal{E}_2^{n+1}(t)\big)^2,\\ &\|D_t\zeta^{n+1}\|_{H^{s-\frac32}}\le\mathcal{A}_2\big(1+\mathcal{E}_2^{n+1}(t)\big),\\ &\|\partial_t\eta^{n+1}\|_{H^{s-\frac32}}\le\mathcal{A}_2\big(1+\mathcal{E}_2^{n+1}(t)\big)^2,\\ &\|\partial_t^2\eta^{n+1}\|_{H^{s-\frac52}}\le\mathcal{A}_4\big(1+\mathcal{E}_2^{n+1}(t)\big)^3.\end{aligned}$$

PROOF. The first two inequalities are obvious. Recall that

$$\begin{aligned}D_t\zeta^{n+1}=&T_{\lambda^n}(V^{n+1}+T_{\zeta^n}B^{n+1})+(T_{V_1^n}-V_1^n)\cdot\zeta^n+[T_{\zeta^n},T_{\lambda^n}]B_1^n\\ &+(\zeta^n-T_{\zeta^n})T_{\lambda^n}B^n+R(\eta_1^n)V_1^n+\zeta^nR(\eta_1^n)B_1^n+R_\omega^n.\end{aligned}$$

Then it follows from Proposition 2.6 and Proposition 5.2 that

$$\|D_t\zeta^{n+1}\|_{H^{s-\frac32}}\le\mathcal{A}_1\mathcal{E}_2^{n+1}(t)+\mathcal{A}_2.$$

For $\partial_t\eta^{n+1}$, we get by the elliptic estimate that

$$\begin{aligned}\|\partial_t\eta^{n+1}\|_{H^{s-\frac32}}\le&C\big(\|\partial_t\zeta^{n+1}\|_{H^{s-\frac52}}+\|\partial_t\eta_1^{n+1}\|_{H^{s-\frac72}}\big)\\ \le&\mathcal{A}_2+\mathcal{A}_1\mathcal{E}_2^{n+1}(t)+C\mathcal{E}_2^{n+1}(t)^2.\end{aligned}$$

Using Proposition 2.6 and Proposition 11.7, we can also deduce that

$$\|\partial_tD_t\zeta^{n+1}\|_{H^{s-\frac52}}\le\mathcal{A}_4\mathcal{E}_2^{n+1}(t)+\mathcal{A}_4,$$

which implies the estimate for $\partial_t^2\eta^{n+1}$. □

LEMMA 11.4. *It holds that*

$$\begin{aligned}&\|(D_tV^{n+1},\widetilde{D}_tV_1^{n+1})\|_{H^{s-\frac12}}\le\mathcal{A}_3\big(1+\mathcal{E}^{n+1}(t)\big),\\ &\|(D_tB^{n+1},\widetilde{D}_tB_1^{n+1},\widetilde{D}_t^bV_{b,1}^{n+1})\|_{H^{s-\frac12}}\le\mathcal{A}_3,\\ &\|(D_t^bV_b^{n+1},\widetilde{D}_t^bV_{b,1}^{n+1})\|_{H^{s-\frac32}}\le\mathcal{A}_2,\\ &\|(\widetilde{D}_tV_1^{n+1},\widetilde{D}_tB_1^{n+1})\|_{H^{s-\frac32}}\le\mathcal{A}_2\big(1+\mathcal{E}_2^{n+1}(t)\big),\\ &\|(\partial_t\widetilde{D}_t(V_1^{n+1},B_1^{n+1}),\partial_t\widetilde{D}_t^bV_{b,1}^{n+1})\|_{H^{s-\frac32}}\le\mathcal{A}_5\big(1+\mathcal{E}_2^{n+1}(t)\big).\end{aligned}$$

PROOF. Recall that $D_t V^{n+1} = \widetilde{D}_t V_1^{n+1}$ and

$$D_t V^{n+1} = -T_{\zeta^n} a^n - T_{a^n}\zeta^{n+1} - R(\zeta^n, a^n) + (T_{V_1^n} - V_1^n)\cdot\nabla V^n.$$

It follows from Lemma 2.10 and Lemma 11.1 that

$$\|(D_t V^{n+1}, \widetilde{D}_t V_1^{n+1})\|_{H^{s-\frac{1}{2}}} \le \mathcal{A}_3 + \mathcal{A}_2\|\zeta^{n+1}\|_{H^{s-\frac{1}{2}}} \le \mathcal{A}_3 + \mathcal{A}_2\mathcal{E}^{n+1}(t).$$

The proof of the other inequalities is similar. □

11.3. Estimate of the velocity

In the sequel, we assume that (H8) holds for $(\eta^{n+1}, \eta_1^{n+1})$, i.e.,

$$\eta^{n+1}(t,x) + 1 \ge \frac{h_0}{2}, \quad \eta_1^{n+1}(t,x) + 1 \ge \frac{h_0}{2}.$$

Recall that v^{n+1} satisfies

$$\begin{cases} -\Delta_{x,y} v^{n+1} = \nabla_{x,y}\times\omega^{n+1} & \text{in } \Omega_t^{n+1}, \\ v^{n+1}|_{y=\eta_1^{n+1}} = (V^{n+1}, B^{n+1}), & v^{n+1}|_{y=-1} = (V_b^{n+1}, 0), \end{cases}$$

and v_1^{n+1} satisfies

$$\begin{cases} -\Delta_{x,y} {v_1}^{n+1} = \nabla_{x,y}\times\omega^{n+1} & \text{in } \Omega_t^{n+1}, \\ v_1^{n+1}|_{y=\eta_1^{n+1}} = (V_1^{n+1}, B_1^{n+1}), & v_1^{n+1}|_{y=-1} = (V_{b,1}^{n+1}, 0). \end{cases}$$

First of all, we apply Proposition 4.10 to obtain

$$\begin{aligned} \|\nabla_{x,z}\widetilde{v^{n+1}}\|_{H^{s-1}(\mathcal{S})} \le& P_{s-\frac{1}{2},\eta_1^{n+1}}\big(\|(V^{n+1}, B^{n+1}, V_b^{n+1})\|_{H^{s-\frac{1}{2}}} + \|\widetilde{\omega^{n+1}}\|_{H^{s-1}}\big) \\ \le& \mathcal{P}(\mathcal{E}_2^{n+1}(t)). \end{aligned} \tag{11.2}$$

Similarly, we have

$$\|\nabla_{x,z}\widetilde{v_1^{n+1}}\|_{H^{s-1}(\mathcal{S})} \le \mathcal{P}(\mathcal{E}_2^{n+1}(t)). \tag{11.3}$$

A direct calculation gives

$$\begin{cases} \Delta_{x,y}\widetilde{\mathcal{D}}_t v_1^{n+1} = h_\omega^{n+1} & \text{in } \Omega_t^{n+1}, \\ \widetilde{\mathcal{D}}_t v_1^{n+1}|_{y=\eta_1^{n+1}} = (\widetilde{D}_t V_1^{n+1}, \widetilde{D}_t B_1^{n+1}), & \widetilde{\mathcal{D}}_t v_1^{n+1}|_{y=-1} = (\widetilde{D}_t^b V_{b,1}^{n+1}, 0), \end{cases}$$

where

$$\begin{aligned} -h_\omega^{n+1} =& \nabla_{x,y}\times\widetilde{\mathcal{D}}_t\omega^{n+1} - \nabla_{x,y} v_2^{n+1}\cdot\nabla_{x,y} w^{n+1} + \Delta_{x,y} v_2^{n+1}\cdot\nabla_{x,y} v_1^{n+1} \\ &+ 2\partial_i v_2^{n+1}\cdot\nabla_{x,y}\partial_i v_1^{n+1}. \end{aligned}$$

We can deduce from Lemma 2.13, (11.2) and (11.3) that

$$\begin{aligned} \|\widetilde{h_\omega^{n+1}}\|_{H^{s-2}(\mathcal{S})} \le& P_{s-\frac{1}{2},\eta_1^{n+1}}\big(\|\widetilde{\widetilde{\mathcal{D}}_t\omega^{n+1}}\|_{H^{s-1}(\mathcal{S})} + \|\widetilde{\omega^{n+1}}\|^2_{H^{s-1}(\mathcal{S})} \\ &+ \|\nabla_{x,z}(\widetilde{v^{n+1}}, \widetilde{v_1^{n+1}})\|^2_{H^{s-1}(\mathcal{S})}\big) \\ \le& P_{s-\frac{1}{2},\eta_1^{n+1}}\big(\|\widetilde{\omega^{n+1}}\|^2_{H^{s-1}(\mathcal{S})} + \|\nabla_{x,z}(\widetilde{v^{n+1}}, \widetilde{v_1^{n+1}})\|^2_{H^{s-1}(\mathcal{S})}\big) \\ \le& \mathcal{P}(\mathcal{E}_2^{n+1}(t)). \end{aligned}$$

While, we know from Lemma 11.4 that

$$\|(\widetilde{D}_t V_1^{n+1}, \widetilde{D}_t B_1^{n+1}, \widetilde{D}_t^b V_{b,1}^{n+1})\|_{H^{s-\frac{1}{2}}} \le \mathcal{A}_3\big(1 + \mathcal{E}^{n+1}(t)\big).$$

Then we apply Proposition 4.10 again to obtain

$$\|\nabla_{x,z}\widetilde{\widetilde{\mathcal{D}}_t v_1^{n+1}}\|_{H^{s-1}(\overline{\mathcal{S}})} \le \mathcal{P}(\mathcal{A}_3, \mathcal{E}^{n+1}(t)). \tag{11.4}$$

Similarly, we have

$$\|\nabla_{x,z}\widetilde{\widetilde{\mathcal{D}}_t v^{n+1}}\|_{H^{s-2}(\overline{\mathcal{S}})} \le \mathcal{P}(\mathcal{A}_3, \mathcal{E}^{n+1}(t)) \tag{11.5}$$

by noting that

$$\|(\widetilde{D}_t V^{n+1}, \widetilde{D}_t B^{n+1}, \widetilde{D}_t^b V_b^{n+1})\|_{H^{s-\frac{3}{2}}} \le \mathcal{P}(\mathcal{A}_3, \mathcal{E}^{n+1}(t)).$$

Using (11.2)-(11.4), we can deduce that

LEMMA 11.5. *It holds that*

$$\|\widetilde{\widetilde{\mathcal{D}}_t \omega^{n+1}}\|_{H^{s-1}(\overline{\mathcal{S}})} \le \mathcal{P}(\mathcal{E}_2^{n+1}(t)),$$
$$\|\widetilde{\widetilde{\mathcal{D}}_t^2 \omega^{n+1}}\|_{H^{s-1}(\overline{\mathcal{S}})} \le \mathcal{P}(\mathcal{A}_3, \mathcal{E}^{n+1}(t)).$$

For v_ω^{n+1} defined by

$$\begin{cases} -\Delta_{x,y} v_\omega^{n+1} = \nabla_{x,y} \times \omega^{n+1} \quad \text{in} \quad \Omega_t^{n+1}, \\ v_\omega^{n+1}|_{y=\eta_1^{n+1}} = 0, \quad v_\omega^{n+1}|_{y=-1} = (V_{b,1}^{n+1}, 0), \end{cases}$$

we can deduce in a similar way that

$$\|\widetilde{v_\omega^{n+1}}\|_{H^s(\overline{\mathcal{S}})} \le \mathcal{P}(\mathcal{E}_2^{n+1}(t)), \tag{11.6}$$

$$\|\widetilde{\widetilde{\mathcal{D}}_t v_\omega^{n+1}}\|_{H^s(\overline{\mathcal{S}})} \le \mathcal{P}(\mathcal{A}_3, \mathcal{E}^{n+1}(t)), \tag{11.7}$$

$$\|\widetilde{\widetilde{\mathcal{D}}_t v_\omega^{n+1}}\|_{X^{s-\frac{1}{2}}([-\frac{1}{2},0])} \le \mathcal{P}(\mathcal{A}_2, \mathcal{E}_2^{n+1}(t)), \tag{11.8}$$

by using the fact that

$$\|\widetilde{D}_t^b V_{b,1}^{n+1}\|_{H^{s-\frac{1}{2}}} \le \mathcal{A}_3, \quad \|\widetilde{D}_t^b V_{b,1}^{n+1}\|_{H^{s-\frac{3}{2}}} \le \mathcal{A}_2.$$

On the other hand, we know that

$$\begin{cases} \Delta_{x,y}\widetilde{\mathcal{D}}_t^2 v_\omega^{n+1} = \widetilde{\mathcal{D}}_t h_\omega^{n+1} + [\Delta_{x,y}, \widetilde{\mathcal{D}}_t]\widetilde{\mathcal{D}}_t v_\omega^{n+1} \quad \text{in} \quad \Omega_t^{n+1}, \\ \widetilde{\mathcal{D}}_t^2 v_\omega^{n+1}|_{y=\eta_1^{n+1}} = 0, \quad \widetilde{\mathcal{D}}_t^2 v_\omega^{n+1}|_{y=-1} = ((\widetilde{D}_t^b)^2 V_{b,1}^{n+1}, 0), \end{cases}$$

where

$$\begin{aligned} h_\omega^{n+1} =& \nabla_{x,y} \times \widetilde{\mathcal{D}}_t \omega^{n+1} - \nabla_{x,y} v_2^{n+1} \cdot \nabla_{x,y}\omega^{n+1} - \nabla_{x,y}\omega^{n+1} \cdot \nabla_{x,y} v_\omega^{n+1} \\ &+ 2\partial_i v_2^{n+1} \cdot \nabla_{x,y}\partial_i v_\omega^{n+1}. \end{aligned}$$

Using (11.4)-(11.7) and Lemma 11.5, we can deduce from Lemma 2.13 that

$$\|[\Delta_{x,y}, \widetilde{\mathcal{D}}_t]\widetilde{\mathcal{D}}_t v_\omega^{n+1}\|_{H^{s-2}(\overline{\mathcal{S}})} + \|\widetilde{\widetilde{\mathcal{D}}_t h_\omega^{n+1}}\|_{H^{s-2}(\overline{\mathcal{S}})} \le \mathcal{P}(\mathcal{A}_3, \mathcal{E}^{n+1}(t)).$$

And by Lemma 11.4, we have

$$\begin{aligned} \|(\widetilde{D}_t^b)^2 V_{b,1}^{n+1}\|_{H^{s-\frac{3}{2}}} \le& \|\partial_t \widetilde{D}_t^b V_{b,1}^{n+1}\|_{H^{s-\frac{3}{2}}} + \|V_b^{n+1}\|_{H^{s-\frac{3}{2}}}\|\nabla^2 P^n(\cdot,-1)\|_{H^{s-\frac{3}{2}}} \\ \le& \mathcal{P}(\mathcal{A}_5, \mathcal{E}^{n+1}(t)). \end{aligned}$$

Then we can prove that

$$\|\nabla_{x,y}\widetilde{\mathcal{D}}_t^2 v_\omega^{n+1}\|_{L^2(\Omega_t^{n+1})} \le \mathcal{P}(\mathcal{A}_5, \mathcal{E}^{n+1}(t)).$$

Thus, Proposition 4.9 ensures that

$$\|\nabla_{x,z}\widetilde{\widetilde{\mathcal{D}}_t^2 v_\omega^{n+1}}\|_{X^{s-\frac{3}{2}}([-\frac{1}{2},0])} \le \mathcal{P}(\mathcal{A}_5,\mathcal{E}^{n+1}(t)). \tag{11.9}$$

11.4. Estimate of the pressure

Recall that the pressure P^{n+1} satisfies

$$\begin{cases} -\Delta_{x,y}P^{n+1} = \big(\partial_i(v_1^j)^{n+1}\partial_j(v_1^i)^{n+1}\big)\circ\Phi_1^{n+1}\circ(\Phi^{n+1})^{-1} \triangleq F, \\ P^{n+1}\mid_{y=\eta^{n+1}}=0, \quad (\partial_y P^{n+1})\mid_{y=-1}=-1. \end{cases}$$

Let $P_1^{n+1} = P^{n+1}+y$, $I_1=[a,0]$ for $a\in(-1,0)$. First of all, we get by a similar proof of Lemma 7.1 that

$$\|\nabla_{x,y}P_1\|_{L^2(\widetilde{\Omega}_t^{n+1})} \le P_{s-\frac{1}{2},\eta^{n+1}}\|\nabla_{x,z}\widetilde{v_1^{n+1}}\|^2_{H^{s-1}(\overline{\mathcal{S}})},$$

from which and Proposition 4.9, it follows that

$$\begin{aligned} \|\nabla_{x,z}\widetilde{P_1^{n+1}}\|_{X^{s-\frac{1}{2}}(I_1)} \le & P_{s+\frac{1}{2},\eta^{n+1}}\big(\|\nabla_{x,y}P_1^{n+1}\|_{L^2(\widetilde{\Omega}_t^{n+1})} \\ & + P_{s-\frac{1}{2},\eta_1^{n+1}}\|\nabla_{x,z}\widetilde{v_1^{n+1}}\|^2_{H^{s-1}(\overline{\mathcal{S}})}\big) \\ \le & \mathcal{P}(\mathcal{A}_2,\mathcal{E}^{n+1}(t)). \end{aligned} \tag{11.10}$$

Similarly, we can deduce that

$$\|\nabla_{x,z}\widetilde{P_1^{n+1}}\|_{X^{s-\frac{3}{2}}(I_1)} \le \mathcal{P}(\mathcal{E}_2^{n+1}(t)). \tag{11.11}$$

These ensure that there exists $y_0\in(-1,-1+h_0]$(in fact, one can take $y_0>-1+a$ with a depending on $\|\eta^{n+1}\|_{H^{s-\frac{1}{2}}}$) so that

$$\begin{aligned} \|P_1(\cdot,y_0)\|_{H^{s+\frac{1}{2}}} &\le \mathcal{P}(\mathcal{A}_2,\mathcal{E}^{n+1}(t)), \\ \|P_1(\cdot,y_0)\|_{H^{s-\frac{1}{2}}} &\le \mathcal{P}(\mathcal{E}_2^{n+1}(t)). \end{aligned}$$

Then by the elliptic estimate in the flat strip, we obtain

$$\|\nabla_{x,y}P_1^{n+1}\|_{H^s(\mathbf{R}^d\times[-1,y_0])} \le \mathcal{P}(\mathcal{A}_2,\mathcal{E}_2^{n+1}(t)), \tag{11.12}$$

$$\|\nabla_{x,y}P_1^{n+1}\|_{H^{s-1}(\mathbf{R}^d\times[-1,y_0])} \le \mathcal{P}(\mathcal{E}_2^{n+1}(t)). \tag{11.13}$$

Using Lemma 11.3, (11.3) and (11.4), we can deduce that

$$\|\widetilde{\mathcal{D}_t F}\|_{H^{s-2}(\overline{\mathcal{S}})} \le \mathcal{P}(\mathcal{A}_3,\mathcal{E}^{n+1}(t)).$$

It is easy to show that

$$\|\nabla_{x,y}\mathcal{D}_t P^{n+1}\|_{L^2(\widetilde{\Omega}_t^{n+1})} \le \mathcal{P}(\mathcal{A}_3,\mathcal{E}^{n+1}(t)).$$

Thus, we can deduce that

$$\|\nabla_{x,z}\widetilde{\mathcal{D}_t P^{n+1}}\|_{X^{s-\frac{3}{2}}(I_1)} + \|\nabla_{x,y}\mathcal{D}_t P^{n+1}(\cdot,-1)\|_{H^{s-\frac{3}{2}}} \le \mathcal{P}(\mathcal{A}_3,\mathcal{E}^{n+1}(t)). \tag{11.14}$$

It follows from (11.10), (11.11) and (11.14) that

LEMMA 11.6. *It holds that*

$$\|a^{n+1}\|_{H^{s-\frac{3}{2}}} \le \mathcal{P}(\mathcal{E}_2^{n+1}(t)),$$
$$\|a^{n+1}\|_{H^{s-\frac{1}{2}}} \le \mathcal{P}(\mathcal{A}_2, \mathcal{E}^{n+1}(t)),$$
$$\|\partial_t a^{n+1}\|_{H^{s-\frac{3}{2}}} \le \mathcal{P}(\mathcal{A}_3, \mathcal{E}^{n+1}(t)).$$

11.5. Estimates of the remainder of DN operator

In this subsection, we establish some estimates for the material derivatives of the remainder of DN operator, which will be used to estimate R_ω^n. For this purpose, we assume that η is a solution of the following equation

$$\partial_t \eta + V \cdot \nabla \eta = B,$$

where $(V, B) = v|_{y=\eta}$. Let $\overline{D}_t \triangleq \partial_t + V \cdot \nabla$ and $\mathcal{D}_t \triangleq \partial_t + v \cdot \nabla_{x,y}$.

Now, we state the main result in this subsection:

PROPOSITION 11.7. *Assume that* $\eta, \overline{D}_t\eta, \overline{D}_t^2\eta \in H^{s-\frac{1}{2}}(\mathbf{R}^d)$ *for* $s > \frac{d}{2}+5$. *Then it holds that*

$$\|\overline{D}_t R(\eta) f\|_{H^{s-\frac{3}{2}}} \le P_{s-\frac{1}{2},\eta,\overline{D}_t\eta} \mathcal{V}_1(t)\big(\|f\|_{H^{s-\frac{1}{2}}} + \|\overline{D}_t f\|_{H^{s-\frac{3}{2}}}\big),$$
$$\|\overline{D}_t^2 R(\eta) f\|_{H^{s-\frac{3}{2}}} \le P_{s-\frac{1}{2},\eta,\overline{D}_t\eta,\overline{D}_t^2\eta} \mathcal{V}_2(t)\big(\|f\|_{H^{s-\frac{1}{2}}} + \|\overline{D}_t f\|_{H^{s-\frac{1}{2}}} + \|\overline{D}_t^2 f\|_{H^{s-\frac{3}{2}}}\big),$$

where $\mathcal{V}_1(t)$ *and* $\mathcal{V}_2(t)$ *are given by*

$$\mathcal{V}_1(t) \triangleq \mathcal{P}\big(\|V(t)\|_{H^{s-\frac{1}{2}}}, \|\nabla_{x,z}\widetilde{v}(t)\|_{H^{s-1}(\overline{\mathcal{S}})}\big),$$
$$\mathcal{V}_2(t) \triangleq \mathcal{P}\big(\|\nabla_{x,z}(\widetilde{v}, \widetilde{\mathcal{D}_t v})(t)\|_{H^{s-1}(\overline{\mathcal{S}})}, \|(V, \overline{D}_t V)\|_{H^{s-\frac{1}{2}}}\big)$$

Let $\phi(t,x,y)$ be a solution of the elliptic equation

$$\begin{cases} \Delta_{x,y}\phi = 0 & \text{in} \quad \Omega_t = \{(x,y) : -1 < y < \eta(t,x)\}, \\ \phi|_{y=\eta(t,x)} = f, & \phi|_{y=-1} = 0. \end{cases}$$

Proposition 4.10 ensures that

$$\|\widetilde{\phi}\|_{H^s(\overline{\mathcal{S}})} \le P_{s-\frac{1}{2},\eta}\|f\|_{H^{s-\frac{1}{2}}}. \tag{11.15}$$

We next establish some estimates for $\mathcal{D}_t^k\phi$ for $k = 1, 2$.

LEMMA 11.8. *It holds that*

$$\|\widetilde{\mathcal{D}_t\phi}\|_{H^s(\overline{\mathcal{S}})} \le P_{s-\frac{1}{2},\eta}\big(\|\overline{D}_t f\|_{H^{s-\frac{1}{2}}} + \|\nabla_{x,z}\widetilde{v}\|_{H^{s-1}(\overline{\mathcal{S}})}\|f\|_{H^{s-\frac{1}{2}}}\big),$$
$$\|\widetilde{\mathcal{D}_t^2\phi}\|_{H^{s-1}(\overline{\mathcal{S}})} \le P_{s-\frac{1}{2},\eta}\mathcal{V}(t)\big(\|\overline{D}_t^2 f\|_{H^{s-\frac{3}{2}}} + \|\overline{D}_t f\|_{H^{s-\frac{3}{2}}} + \|f\|_{H^{s-\frac{3}{2}}}\big),$$
$$\|(\partial_z - T_A)\widetilde{\mathcal{D}_t\phi}\|_{X^{s-\frac{3}{2}}([-\frac{1}{2},0])} \le P_{s-\frac{1}{2},\eta}\big(\|\overline{D}_t f\|_{H^{s-\frac{3}{2}}} + \|\nabla_{x,z}\widetilde{v}(t)\|_{H^{s-1}(\overline{\mathcal{S}})}\|f\|_{H^{s-\frac{1}{2}}}\big),$$
$$\|(\partial_z - T_A)\widetilde{\mathcal{D}_t^2\phi}\|_{X^{s-\frac{3}{2}}([-\frac{1}{2},0])} \le P_{s-\frac{1}{2},\eta}\mathcal{V}(t)\big(\|\overline{D}_t^2 f\|_{H^{s-\frac{3}{2}}} + \|\overline{D}_t f\|_{H^{s-\frac{1}{2}}} + \|f\|_{H^{s-\frac{1}{2}}}\big).$$

Here $\mathcal{V}(t) \triangleq \mathcal{P}\big(\|\nabla_{x,z}\widetilde{v}(t)\|_{H^{s-1}(\overline{\mathcal{S}})}, \|\nabla_{x,z}\widetilde{\mathcal{D}_t v}(t)\|_{H^{s-1}(\overline{\mathcal{S}})}\big)$.

PROOF. A direct calculations gives

$$\begin{cases} \Delta_{x,y}\mathcal{D}_t\phi = \Delta_{x,y}v \cdot \nabla_{x,y}\phi + 2\nabla_{x,y}v \cdot \nabla_{x,y}^2\phi \triangleq F, \\ \mathcal{D}_t\phi|_{y=\eta} = \overline{D}_t f, \quad \mathcal{D}_t\phi|_{y=-1} = 0. \end{cases}$$

By Lemma 2.13, it is easy to show that

$$\|\widetilde{F}\|_{H^{s-2}(\mathcal{S})} \leq P_{s-\frac{1}{2},\eta}\|\nabla_{x,z}\widetilde{v}\|_{H^{s-1}(\mathcal{S})}\|\widetilde{\phi}\|_{H^s(\mathcal{S})}.$$

Then we deduce from Proposition 4.10 and (11.15) that

$$\|\widetilde{\mathcal{D}_t\phi}\|_{H^s(\mathcal{S})} \leq P_{s-\frac{1}{2},\eta}\big(\|\overline{D}_t f\|_{H^{s-\frac{1}{2}}} + \|\nabla_{x,z}\widetilde{v}\|_{H^{s-1}(\mathcal{S})}\|f\|_{H^{s-\frac{1}{2}}}\big). \tag{11.16}$$

Similar, we have

$$\|\widetilde{\mathcal{D}_t\phi}\|_{H^{s-1}(\mathcal{S})} \leq P_{s-\frac{1}{2},\eta}\big(\|\overline{D}_t f\|_{H^{s-\frac{3}{2}}} + \|\nabla_{x,z}\widetilde{v}\|_{H^{s-2}(\mathcal{S})}\|f\|_{H^{s-\frac{3}{2}}}\big).$$

Then a similar argument of Proposition 5.2 implies the third inequality.

By Lemma 2.13 again, we have

$$\begin{aligned}\|\widetilde{\mathcal{D}_t F}\|_{H^{s-3}(\mathcal{S})} \leq & P_{s-\frac{1}{2},\eta}\|\nabla_{x,z}\widetilde{v}\|^2_{H^{s-2}(\mathcal{S})}\|\widetilde{\mathcal{D}_t\phi}\|_{H^{s-1}(\mathcal{S})} \\ & + P_{s-\frac{1}{2},\eta}\big(\|\nabla_{x,z}\widetilde{\mathcal{D}_t v}\|_{H^{s-2}(\mathcal{S})} + \|\nabla_{x,z}\widetilde{v}\|^2_{H^{s-2}(\mathcal{S})}\big)\|\widetilde{\phi}\|_{H^{s-1}(\mathcal{S})}.\end{aligned}$$

This implies the second inequality. We also have

$$\begin{aligned}\|\widetilde{\mathcal{D}_t F}\|_{H^{s-2}(\mathcal{S})} \leq & P_{s-\frac{1}{2},\eta}\|\nabla_{x,z}\widetilde{v}\|^2_{H^{s-1}(\mathcal{S})}\|\widetilde{\mathcal{D}_t\phi}\|_{H^s(\mathcal{S})} \\ & + P_{s-\frac{1}{2},\eta}\big(\|\nabla_{x,z}\widetilde{\mathcal{D}_t v}\|_{H^{s-1}(\mathcal{S})} + \|\nabla_{x,z}\widetilde{v}\|^2_{H^{s-1}(\mathcal{S})}\big)\|\widetilde{\phi}\|_{H^s(\mathcal{S})} \\ \leq & P_{s-\frac{1}{2},\eta}\|\nabla_{x,z}\widetilde{v}\|^2_{H^{s-1}(\mathcal{S})}\big(\|\overline{D}_t f\|_{H^{s-\frac{1}{2}}} + \|\nabla_{x,z}\widetilde{v}\|_{H^{s-1}(\mathcal{S})}\|f\|_{H^{s-\frac{1}{2}}}\big) \\ & + P_{s-\frac{1}{2},\eta}\big(\|\nabla_{x,z}\widetilde{\mathcal{D}_t v}\|_{H^{s-1}(\mathcal{S})} + \|\nabla_{x,z}\widetilde{v}\|^2_{H^{s-1}(\mathcal{S})}\big)\|f\|_{H^{s-\frac{1}{2}}}.\end{aligned}$$

Then a similar argument of Proposition 5.2 implies the last inequality. □

LEMMA 11.9. *It holds that*

$$\begin{aligned}\|\overline{D}_t\widetilde{\phi}|_{z=0}\|_{H^{s-\frac{3}{2}}} &\leq P_{s-\frac{1}{2},\eta}\big(\|\nabla_{x,z}\widetilde{v}\|_{H^{s-1}(\mathcal{S})}\|f\|_{H^{s-\frac{3}{2}}} + \|\overline{D}_t f\|_{H^{s-\frac{3}{2}}}\big), \\ \|\overline{D}_t^2\widetilde{\phi}|_{z=0}\|_{H^{s-\frac{3}{2}}} &\leq P_{s-\frac{1}{2},\eta}\mathcal{V}(t)\big(\|f\|_{H^{s-\frac{3}{2}}} + \|D_t f\|_{H^{s-\frac{3}{2}}} + \|\overline{D}_t^2 f\|_{H^{s-\frac{3}{2}}}\big).\end{aligned}$$

In addition, we have

$$\begin{aligned}&\big\|\overline{D}_t\nabla_{x,z}\widetilde{\phi}|_{z=0}\big\|_{H^{s-\frac{5}{2}}} \leq P_{s-\frac{1}{2},\eta,\overline{D}_t\eta}\big(\|\nabla_{x,z}\widetilde{v}\|_{H^{s-1}(\mathcal{S})}\|f\|_{H^{s-\frac{3}{2}}} + \|\overline{D}_t f\|_{H^{s-\frac{3}{2}}}\big), \\ &\big\|\overline{D}_t^2\nabla_{x,z}\widetilde{\phi}|_{z=0}\big\|_{H^{s-\frac{5}{2}}} \leq P_{s-\frac{1}{2},\eta,\overline{D}_t\eta,\overline{D}_t^2\eta}\mathcal{V}(t)\big(\|f\|_{H^{s-\frac{3}{2}}} \\ &\quad + \|\overline{D}_t f\|_{H^{s-\frac{3}{2}}} + \|\overline{D}_t^2 f\|_{H^{s-\frac{3}{2}}}\big).\end{aligned}$$

PROOF. The first two inequalities follows from Lemma 11.8 and the fact that

$$\overline{D}_t^k\widetilde{\phi}|_{z=0} = \widetilde{\mathcal{D}_t^k\phi}|_{z=0}.$$

The last two inequalities can be deduced from Lemma 2.11, Lemma 11.8 and the formulas

$$\begin{aligned}\overline{D}_t\nabla\widetilde{\phi}|_{z=0} &= \widetilde{\mathcal{D}_t\nabla\phi}|_{z=0} + \widetilde{\mathcal{D}_t\partial_y\phi}|_{z=0}\nabla\eta + \widetilde{\partial_y\phi}|_{z=0}\overline{D}_t\nabla\eta, \\ \overline{D}_t\partial_z\widetilde{\phi}|_{z=0} &= \widetilde{\mathcal{D}_t\partial_y\phi}|_{z=0}(1+\eta+\delta|D|\eta) + \widetilde{\partial_y\phi}|_{z=0}\overline{D}_t(\eta+\delta|D|\eta).\end{aligned}$$

We omit the details. □

PROOF OF PROPOSITION 11.7. We write

$$\begin{aligned}\overline{D}_t R_1(\eta)f =&\big[\overline{D}_t, T_{\zeta_1}T_A - T_{\zeta_1 A}\big]\widetilde{\phi}|_{z=0} + \big(T_{\zeta_1}T_A - T_{\zeta_1 A}\big)\widetilde{\mathcal{D}_t\phi}|_{z=0}\\ =&\big(T_{\partial_t\zeta_1}T_A + T_{\zeta_1}T_{\partial_t A} - T_{\partial_t(\zeta_1 A)}\big)\widetilde{\phi}|_{z=0} + \big[V\cdot\nabla, T_{\zeta_1}T_A - T_{\zeta_1 A}\big]\widetilde{\phi}|_{z=0}\\ &+ \big(T_{\zeta_1}T_A - T_{\zeta_1 A}\big)\widetilde{\mathcal{D}_t\phi}|_{z=0}.\end{aligned}$$

Using Proposition 2.6 and Lemma 11.9, and the following trick

$$\|\partial_t g\|_{H^{s-\frac{5}{2}}} \le \|\overline{D}_t g\|_{H^{s-\frac{5}{2}}} + \|V\|_{H^{s-\frac{5}{2}}}\|g\|_{H^{s-\frac{3}{2}}},$$

we can deduce that

$$\|\overline{D}_t R_1(\eta)f\|_{H^{s-\frac{3}{2}}} \le P_{s-\frac{1}{2},\eta,\overline{D}_t\eta}\mathcal{V}(t)\big(\|f\|_{H^{s-\frac{3}{2}}} + \|\overline{D}_t f\|_{H^{s-\frac{3}{2}}}\big).$$

Similarly, we have

$$\|\overline{D}_t R_2(\eta)f\|_{H^{s-\frac{3}{2}}} \le P_{s-\frac{1}{2},\eta,\overline{D}_t\eta}\mathcal{V}(t)\big(\|f\|_{H^{s-\frac{1}{2}}} + \|\overline{D}_t f\|_{H^{s-\frac{3}{2}}}\big).$$

The same estimates holds for $\overline{D}_t R_3(\eta)$ by using the following type estimate

$$\begin{aligned}&\|\overline{D}_t R(f,g)\|_{H^{s-\frac{3}{2}}} \le \|R(\partial_t f,g) + R(f,\partial_t g) + V\cdot\nabla R(f,g)\|_{H^{s-\frac{3}{2}}}\\ &\le C\|\partial_t f\|_{H^{s-\frac{5}{2}}}\|g\|_{H^{s-\frac{3}{2}}} + C\|\partial_t g\|_{H^{s-\frac{5}{2}}}\|f\|_{H^{s-\frac{3}{2}}} + C\|V\|_{H^{s-\frac{3}{2}}}\|f\|_{H^{s-\frac{3}{2}}}\|g\|_{H^{s-\frac{3}{2}}}.\end{aligned}$$

The estimates for $\overline{D}_t^2 R(\eta)f$ can be similarly proved by a very tedious computation with the help of Lemma 11.8 and Lemma 11.9. We omit the details. □

Applying Proposition 11.7 with $v = v_1^n$ and $(V,B) = (V_1^n, B_1^n)$, we deduce that

LEMMA 11.10. *It holds that*

$$\begin{aligned}&\|D_t^u R(\eta_1^n)(V_1^n, B_1^n)\|_{H^{s-\frac{3}{2}}} \le \mathcal{P}(E_1, E_2, E_4),\\ &\|(D_t^u)^2 R(\eta_1^n)(V_1^n, B_1^n)\|_{H^{s-\frac{3}{2}}} \le \mathcal{A}_6.\end{aligned}$$

11.6. Energy estimates

PROPOSITION 11.11. *It holds that*

$$\|\widetilde{\omega}^{n+1}(t)\|_{H^{s-1}(\mathcal{S})} \le \|\widetilde{\omega}_0\|_{H^{s-1}(\mathcal{S})} + \int_0^t \mathcal{P}(\mathcal{A}_1, \mathcal{E}^{n+1}(t'))dt'.$$

PROOF. Recall that the vorticity ω^{n+1} satisfies

$$\begin{cases}\partial_t\omega^{n+1} + \big((v^{n+1})^h\cdot\nabla + (v_1^{n+1})^{d+1}\partial_y\big)\omega^{n+1} = \omega^{n+1}\cdot\nabla_{x,y}v_1^{n+1},\\ \omega^{n+1}|_{t=0} = \omega_0.\end{cases}$$

In terms of the (x,z) variable, we have

$$\partial_t\widetilde{\omega}^{n+1} + \overline{v}^{n+1}\cdot\nabla_{x,z}\widetilde{\omega}^{n+1} = F^{n+1}, \tag{11.17}$$

where

$$\overline{v}^{n+1} = \Big((\widetilde{v}^{n+1})^h, \frac{1}{\partial_z\rho_{\delta,\eta_1^{n+1}}}\big(\widetilde{v_1}^{d+1} - \partial_t\rho_{\delta,\eta_1^{n+1}} - (\widetilde{v}^{n+1})^h\cdot\nabla\rho_{\delta,\eta_1^{n+1}}\big)\Big),$$

$$F^{n+1} = (\widetilde{\omega}^{n+1})^h\cdot\Big(\nabla\widetilde{v_1}^{n+1} - \frac{\nabla\rho_{\delta,\eta_1^{n+1}}}{\partial_z\rho_{\delta,\eta_1^{n+1}}}\partial_z\widetilde{v_1}\Big) + (\widetilde{\omega}^{n+1})^{d+1}\frac{\partial_z\widetilde{v_1}}{\partial_z\rho_{\delta,\eta_1^{n+1}}}.$$

By Lemma 2.13, it is easy to see that

$$\|F^{n+1}\|_{H^{s-1}(\mathcal{S})} \le \mathcal{P}_{s-\frac{1}{2},\eta_1^{n+1}}\|\nabla_{x,z}(\widetilde{v^{n+1}}, \widetilde{v_1^{n+1}})\|_{H^{s-1}(\mathcal{S})}\|\widetilde{\omega}^{n+1}\|_{H^{s-1}(\mathcal{S})}.$$

By Lemma 9.6, we have

$$\|\nabla_{x,z}\overline{v}^{n+1}\|_{H^{s-1}(\overline{\mathcal{S}})} \leq \mathcal{P}_{s-\frac{1}{2},\eta_1^{n+1}}\|\nabla_{x,z}(\widetilde{v^{n+1}},\widetilde{v_1^{n+1}})\|_{H^{s-1}(\overline{\mathcal{S}})}.$$

Then by a similar proof of Proposition 9.5, we deduce

$$\frac{d}{dt}\|\widetilde{\omega}^{n+1}\|_{H^{s-1}(\overline{\mathcal{S}})} \leq \mathcal{P}_{s-\frac{1}{2},\eta_1^{n+1}}\|\nabla_{x,z}(\widetilde{v^{n+1}},\widetilde{v_1^{n+1}})\|_{H^{s-1}(\overline{\mathcal{S}})}. \tag{11.18}$$

This together with (11.2) and (11.3) gives the proposition. □

PROPOSITION 11.12. *It holds that*

$$\begin{aligned}&\|(V^{n+1},B^{n+1},V_b^{n+1})(t)\|_{H^{s-\frac{1}{2}}}+\|\zeta^{n+1}(t)\|_{H^{s-\frac{3}{2}}}\\ &\leq\Big(\|(V_0,B_0,V_{b,0})\|_{H^{s-\frac{1}{2}}}+\mathcal{A}_3\int_0^t(1+\mathcal{E}^{n+1}(t'))dt'\Big)e^{t\mathcal{A}_1}.\end{aligned}$$

PROOF. It follows from Lemma 11.4 that

$$\begin{aligned}\frac{1}{2}\frac{d}{dt}\|V^{n+1}\|^2_{H^{s-\frac{1}{2}}} \leq&\mathcal{A}_3(1+\mathcal{E}^{n+1}(t))\|V^{n+1}\|_{H^{s-\frac{1}{2}}}\\ &+\big\langle\langle D\rangle^{s-\frac{1}{2}}T_{V^n}\cdot\nabla V^{n+1},\langle D\rangle^{s-\frac{1}{2}}V^{n+1}\big\rangle.\end{aligned}$$

We write

$$\begin{aligned}&\big\langle\langle D\rangle^{s-\frac{1}{2}}T_{V^n}\cdot\nabla V^{n+1},\langle D\rangle^{s-\frac{1}{2}}V^{n+1}\big\rangle\\ &=\big\langle[\langle D\rangle^{s-\frac{1}{2}},T_{V^n}\cdot\nabla]V^{n+1},\langle D\rangle^{s-\frac{1}{2}}V^{n+1}\big\rangle-\frac{1}{2}\big\langle T_{\nabla\cdot V^n}\langle D\rangle^{s-\frac{1}{2}}V^{n+1},\langle D\rangle^{s-\frac{1}{2}}V^{n+1}\big\rangle,\end{aligned}$$

by Lemma 2.19, which is bounded by

$$C\|V^n\|_{W^{1,\infty}}\|V^{n+1}\|^2_{H^{s-\frac{1}{2}}} \leq C\|V^n\|_{H^{s-\frac{1}{2}}}\|V^{n+1}\|^2_{H^{s-\frac{1}{2}}}.$$

This shows that

$$\frac{d}{dt}\|V^{n+1}\|_{H^{s-\frac{1}{2}}} \leq\mathcal{A}_3(1+\mathcal{E}^{n+1}(t))+C\|V^n\|_{H^{s-\frac{1}{2}}}\|V^{n+1}\|_{H^{s-\frac{1}{2}}}.$$

Then Gronwall's inequality ensures that

$$\|V^{n+1}(t)\|_{H^{s-\frac{1}{2}}} \leq \Big(\|V_0\|_{H^{s-\frac{1}{2}}}+\mathcal{A}_3\int_0^t(1+\mathcal{E}^{n+1}(t'))dt'\Big)e^{t\mathcal{A}_1}.$$

The estimates for the other terms are similar. □

Similarly, we can prove by Lemma 11.4 and Lemma 11.3 that

PROPOSITION 11.13. *It holds that*

$$\begin{aligned}&\|(V_1^{n+1},B_1^{n+1},V_{b,1}^{n+1})(t)\|_{H^{s-\frac{1}{2}}}+\|\eta_1^{n+1}(t)\|_{H^{s-\frac{1}{2}}}\\ &\leq\Big(\|(V_0,B_0,V_{b,0},\eta_0)\|_{H^{s-\frac{1}{2}}}+\mathcal{A}_3\int_0^t(1+\mathcal{E}^{n+1}(t'))dt'\Big)e^{C\int_0^t\mathcal{E}_2^{n+1}(t')dt'}.\end{aligned}$$

Following the proof of Proposition 9.4, we can deduce that

PROPOSITION 11.14. *It holds that*

$$\begin{aligned}\frac{1}{2}\frac{d}{dt}\big(\|D_t\zeta^{n+1}\|^2_{H^{s-1}}+\|T_{\sqrt{a^n\lambda^n}}\zeta^{n+1}\|^2_{H^{s-1}}\big)\leq\mathcal{A}_4\big(\|D_t\zeta_1^{n+1}\|^2_{H^{s-1}}+\|\zeta\|^2_{H^{s-\frac{1}{2}}}\big)\\ +\big\langle(f_1^n+f_2^n+D_tf_3^n+D_tf_4^n+D_tf_\omega^n)_{s-1},(D_t\zeta^{n+1})_{s-1}\big\rangle.\end{aligned}$$

11.7. Nonlinear estimates

Recall that

$$f_1^n = D_t\big((T_{V_1^n} - V_1^n)\cdot\nabla\zeta^n + [T_{\zeta^n}, T_{\lambda^n}]B_1^n\big) + (T_{a^n\lambda^n} - T_{\lambda^n}T_{a^n})\zeta^{n+1}.$$

LEMMA 11.15. *It holds that*

$$\|f_1^n\|_{H^{s-1}} \le \mathcal{A}_5.$$

PROOF. We write

$$D_t(T_{V_1^n} - V_1^n)\cdot\nabla\zeta^n = -[D_t, T_{\nabla\zeta^n}]\cdot V_1^n + T_{\nabla\zeta^n}\cdot D_tV_1^n - D_tR(\nabla\zeta^n, V_1^n).$$

It follows from Proposition 2.21 that

$$\|[D_t, T_{\nabla\zeta^n}]\cdot V_1^n\|_{H^{s-1}} \le C(E_1+E_4)\|V_1^n\|_{H^{s-1}} \le \mathcal{A}_4.$$

And by Lemma 2.10, we get

$$\|T_{\nabla\zeta^n}\cdot D_tV_1^n\|_{H^{s-1}} \le CE_1\|D_tV_1^n\|_{H^{s-1}} \le \mathcal{A}_5.$$

By Lemma 2.10 again, we get

$$\begin{aligned}\|D_tR(\nabla\zeta^n, V_1^n)\|_{H^{s-1}} \le& \|R(\nabla\partial_t\zeta^n, V_1^n)\|_{H^{s-1}} + \|R(\nabla\zeta^n, \partial_tV_1^n)\|_{H^{s-1}}\\ &+ \|T_{V^n}\cdot\nabla R(\nabla\zeta^n, V_1^n)\|_{H^s} \le \mathcal{A}_4.\end{aligned}$$

We write

$$\begin{aligned}D_t[T_{\zeta^n}, T_{\lambda^n}]B_1^n =& [T_{\partial_t\zeta^n}, T_{\lambda^n}]B_1^n + [T_{\zeta^n}, T_{\partial_t\lambda^n}]B_1^n + [T_{\zeta^n}, T_{\lambda^n}]D_tB_1^n\\ &+ \big[T_{V^n}\cdot\nabla, [T_{\zeta^n}, T_{\lambda^n}]\big]B_1^n,\end{aligned}$$

which along with Proposition 2.6 and Corollary 2.9 gives

$$\|D_t[T_{\zeta^n}, T_{\lambda^n}]B_1^n\|_{H^{s-1}} \le \mathcal{A}_4.$$

The proof is finished. □

Recall that

$$f_2^n = [D_t, T_{\lambda^n}]U^{n+1} + T_{\lambda^n}\big(h_1^n + [D_t, T_{\zeta^n}]B^{n+1}\big).$$

LEMMA 11.16. *It holds that*

$$\begin{aligned}&\|f_2^n\|_{H^{s-\frac32}} \le \mathcal{P}(E_1, E_2, E_4)\big(1+\mathcal{E}_2^{n+1}(t)\big),\\ &\|D_tf_2^n\|_{H^{s-\frac32}} \le \mathcal{A}_5\big(1+\mathcal{E}^{n+1}(t)\big).\end{aligned}$$

PROOF. The first inequality is obvious. We turn to the second inequality. It follows from Proposition 2.23 and Lemma 11.2 that

$$\begin{aligned}\|D_t[D_t, T_{\lambda^n}]U^{n+1}\|_{H^{s-\frac32}} \le& \mathcal{A}_5\big(\|U^{n+1}\|_{H^{s-\frac12}} + \|D_tU^{n+1}\|_{H^{s-\frac12}}\big)\\ \le& \mathcal{A}_5\big(1+\mathcal{E}^{n+1}(t)\big).\end{aligned}$$

We write

$$\begin{aligned}D_tT_{\lambda^n}\big(h_1^n + [D_t, T_{\zeta^n}]B^{n+1}\big) =& [D_t, T_{\lambda^n}]\big(h_1^n + [D_t, T_{\zeta^n}]B^{n+1}\big)\\ &+ T_{\lambda^n}\big(D_th_1^n + D_t[D_t, T_{\zeta^n}]B^{n+1}\big),\end{aligned}$$

which along with Proposition 2.21 and Proposition 2.6 gives

$$\begin{aligned}\|D_tT_{\lambda^n}\big(h_1^n + [D_t, T_{\zeta^n}]B^{n+1}\big)\|_{H^{s-\frac32}} \le& \mathcal{A}_5\big(\|h_1^n\|_{H^{s-\frac12}} + \|B^{n+1}\|_{H^{s-\frac12}}\\ &+ \|D_th_1^n\|_{H^{s-\frac12}} + \|D_tB^{n+1}\|_{H^{s-\frac12}}\big).\end{aligned}$$

Similar to Lemma 11.15, we can prove that

$$\|D_t h_1^n\|_{H^{s-\frac{1}{2}}} \le \mathcal{A}_5.$$

This together with Lemma 11.4 gives

$$\|D_t T_{\lambda^n}\big(h_1^n + [D_t, T_{\zeta^n}]B^{n+1}\big)\|_{H^{s-\frac{2}{3}}} \le \mathcal{A}_5\big(1+\mathcal{E}^{n+1}(t)\big).$$

Putting the above estimates together gives the second inequality. □

Recall that

$$\begin{aligned} f_3^n &= (\zeta^n - T_{\zeta^n})T_{\lambda^n}B^n + (\zeta^n - T_{\zeta^n})R(\eta_1^n)B_1^n, \\ f_4^n &= R(\eta_1^n)V_1^n + T_{\zeta^n}R(\eta_1^n)B_1^n. \end{aligned}$$

Then we can deduce from Proposition 5.2 and Lemma 11.10 that

LEMMA 11.17. *It holds that*

$$\begin{aligned} &\|f_3^n\|_{H^{s-1}} \le \mathcal{A}_1\|\eta^n\|_{H^s}, \quad \|f_3^n\|_{H^{s-\frac{1}{2}}} \le \mathcal{A}_3, \\ &\|D_t^u f_4^n\|_{H^{s-\frac{3}{2}}} \le \mathcal{P}(E_1,E_2,E_4), \quad \|(D_t^u)^2 f_4^n\|_{H^{s-\frac{3}{2}}} \le \mathcal{A}_6. \end{aligned}$$

Finally, we have

LEMMA 11.18. *It holds that*

$$\|D_t^u f_\omega^n\|_{H^{s-\frac{3}{2}}} \le \mathcal{P}(E_1,E_2,E_4), \quad \|(D_t^u)^2 f_\omega^n\|_{H^{s-\frac{3}{2}}} \le \mathcal{A}_6.$$

11.8. Completion of the uniform estimate

Let

$$\mathcal{E}_0 \triangleq \|\widetilde{\omega_0}\|_{H^{s-1}(\mathcal{S})} + \|\eta_0\|_{H^{s+\frac{1}{2}}} + \|\widetilde{v_0}\|_{H^{s+\frac{1}{2}}(\mathcal{S})} + \|D_t\zeta(0)\|_{H^{s-1}} + \|T_{\sqrt{a_0\lambda_0}}\zeta_0\|_{H^{s-1}},$$

which is bounded by $\mathcal{P}(\|\eta_0\|_{H^{s+\frac{1}{2}}})\big(\|v_0\|_{H^{s+\frac{1}{2}}(\Omega_0)} + \|\eta_0\|_{H^{s+\frac{1}{2}}}\big)$.

Let us first assume that there exists a maximal time $T_n \in (0,T]$ so that the solution satisfies

$$\mathcal{E}^k(t) \le \mathcal{P}_0, \quad \mathcal{E}_2^k(t) \le \mathcal{P}_1. \tag{11.19}$$

for $k = 1, \cdots, n+1$ and $t \in [0, T_n)$ and some $\mathcal{P}_0, \mathcal{P}_1$ determined later. Under this assumption, we will verify $(H1) - (H8)$ for the solution in the $(n+1)$-th iteration. Moreover, we will show that

$$\mathcal{E}^{n+1}(t) \le \frac{\mathcal{P}_0}{2}, \quad \mathcal{E}_2^{n+1}(t) \le \frac{\mathcal{P}_1}{2}. \tag{11.20}$$

which implies that $T_n = T$ by a continuous argument.

Obviously, we can take $E_1 = \mathcal{P}_1$. If $(\eta^{n+1}, \eta_1^{n+1})$ satisfies (H8), then we deduce from (11.3), (11.6), Lemma 11.6 and (11.12) that

$$\begin{aligned} &\|\nabla_{x,z}\widetilde{v_1^n}(t)\|_{H^{s-1}(\mathcal{S})} + \|\widetilde{v}_\omega^{n+1}(t)\|_{H^s(\mathcal{S})} + \|a^{n+1}(t)\|_{H^{s-\frac{3}{2}}} \\ &\quad + \|\nabla P^{n+1}(t,\cdot,-1)\|_{H^{s-\frac{3}{2}}} \le \mathcal{P}(\mathcal{E}_2^{n+1}(t)) \le C(\mathcal{P}_1) \triangleq E_2. \end{aligned}$$

This in turn implies that we can take T_1 small enough depending on $h_0, \mathcal{P}_1, E_2$ so that for $t \in \big[0, \min(T_n, T_1)\big)$,

$$\eta^{n+1}(t,x) + 1 \ge \frac{h_0}{2}, \quad \eta_1^{n+1}(t,x) + 1 \ge \frac{h_0}{2}.$$

By Lemma 11.6, Lemma 11.1 and (11.12), we have

$$\|a^{n+1}(t)\|_{H^{s-\frac{1}{2}}} + \|\nabla P^{n+1}(t,\cdot,-1)\|_{H^{s-\frac{3}{2}}} \leq \mathcal{P}(\mathcal{A}_2, \mathcal{E}^{n+1}(t)) \leq C(\mathcal{P}_0, \mathcal{P}_1) \triangleq E_3^1,$$
$$\|\eta^{n+1}(t)\|_{H^{s+\frac{1}{2}}} \leq \mathcal{A}_2 \mathcal{E}^{n+1}(t) \leq \mathcal{A}_2 \mathcal{P}_0 \triangleq E_3^2.$$

Then we deduce from Proposition 11.11, Proposition 11.12 and Proposition 11.13 that

$$\begin{aligned}\mathcal{E}_2^{n+1}(t) \leq& \Big(\mathcal{E}^{n+1}(0) + \mathcal{A}_3 \int_0^t \mathcal{P}(\mathcal{E}^{n+1}(t'))dt'\Big) e^{\mathcal{A}_1 t + C\int_0^t \mathcal{E}^{n+1}(t')dt'}\\ \leq& C_1\big(\mathcal{E}_0 + tC(\mathcal{P}_0,\mathcal{P}_1)\big)e^{tC(\mathcal{P}_0,\mathcal{P}_1)},\end{aligned}$$

which ensures that we can take $\mathcal{P}_1 = 2C_1\mathcal{E}_0$ and T_2 small enough depending on $\mathcal{E}_0, \mathcal{P}_0$ so that for $t \in \big[0, \min(T_n, T_1, T_2)\big)$,

$$\mathcal{E}_2^{n+1}(t) \leq \frac{\mathcal{P}_1}{2}.$$

By Lemma 11.4, Lemma 11.3, Lemma 11.5 and (11.8), we have

$$\begin{aligned}&\|D_t(V_1^{n+1}, B_1^{n+1})(t)\|_{H^{s-\frac{3}{2}}} + \|\partial_t \eta^{n+1}(t)\|_{H^{s-\frac{1}{2}}} + \|\widetilde{D}_t^u \eta_1^{n+1}(t)\|_{H^{s-\frac{1}{2}}}\\ &+ \|\widetilde{\mathcal{D}_t \omega^{n+1}}(t)\|_{H^{s-1}(\mathcal{S})} + \|\widetilde{\mathcal{D}_t v_\omega^{n+1}}(t)\|_{X^{s-\frac{1}{2}}([-\frac{1}{2},0])}\\ &\leq \mathcal{P}(\mathcal{A}_2, \mathcal{E}_2^{n+1}(t)) \leq C(\mathcal{P}_1) \triangleq E_4.\end{aligned}$$

By Lemma 11.4, Lemma 11.3, Lemma 11.6, (11.4) and (11.14), we obtain

$$\begin{aligned}&\|\big(D_t(V_1^{n+1}, B_1^{n+1}), D_t^b V_{b,1}^{n+1}\big)(t)\|_{H^{s-\frac{1}{2}}} + \|\partial_t^2 \eta^{n+1}(t)\|_{H^{s-\frac{5}{2}}} + \|\partial_t a^{n+1}(t)\|_{H^{s-\frac{3}{2}}}\\ &+ \|\widetilde{\mathcal{D}_t^2 \omega^n}(t)\|_{H^{s-1}(\mathcal{S})} + \|\nabla_{x,z}\widetilde{\mathcal{D}_t v_1^{n+1}}(t)\|_{H^{s-1}(\mathcal{S})} + \|\nabla \partial_t P^{n+1}(t,\cdot,-1)\|_{H^{s-\frac{3}{2}}}\\ &+ \|(\widetilde{D}_t^u)^2 \eta_1^{n+1}(t)\|_{H^{s-\frac{1}{2}}} \leq \mathcal{P}(\mathcal{A}_4, \mathcal{E}^{n+1}(t)) \leq C(\mathcal{P}_0, \mathcal{P}_1) \triangleq E_5.\end{aligned}$$

While, (H5) and the initial conditions (1.7) ensure that there exists T_3 small enough depending on E_5 so that (H7) holds for $t \in \big[0, \min(T_n, T_1, T_2, T_3)\big)$.

By Lemma 11.4, Lemma 11.5 and (11.9), we have

$$\begin{aligned}&\|\partial_t\big(D_t(V_1^{n+1}, B_1^{n+1}), D_t^b V_{b,1}^{n+1}\big)(t)\|_{H^{s-\frac{3}{2}}} + \|\widetilde{\mathcal{D}_t^2 v_\omega^{n+1}}(t)\|_{H^{s-1}(\mathcal{S})}\\ &\leq \mathcal{P}(\mathcal{A}_5, \mathcal{E}^{n+1}(t)) \leq \mathcal{P}(\mathcal{A}_5, \mathcal{P}_0) \triangleq E_6.\end{aligned}$$

To complete the proof, it remains to prove the first inequality of (11.20). We know from Proposition 11.14 that

$$\begin{aligned}\frac{1}{2}\frac{d}{dt}\big(\|D_t\zeta^{n+1}\|_{H^{s-1}}^2 + \|T_{\sqrt{a^n\lambda^n}}\zeta^{n+1}\|_{H^{s-1}}^2\big) \leq& \mathcal{A}_4\big(\|D_t\zeta^{n+1}\|_{H^{s-1}}^2 + \|\zeta\|_{H^{s-\frac{1}{2}}}^2\big)\\ &+ \big\langle (f_1^n + f_2^n + D_t f_3^n + D_t f_4^n + f_\omega^n)_{s-1}, (D_t\zeta^{n+1})_{s-1}\big\rangle.\end{aligned}$$

By Lemma 11.15, we have

$$\big\langle (f_1^n)_{s-1}, (D_t\zeta^{n+1})_{s-1}\big\rangle \leq \mathcal{A}_5 \|D_t\zeta^{n+1}\|_{H^{s-1}}.$$

We write

$$\begin{aligned}\big\langle (f_2^n)_{s-1}, (D_t\zeta^{n+1})_{s-1}\big\rangle =& \big\langle (f_2^n)_{s-1}, [\langle D\rangle^s, D_t]\zeta^{n+1})\big\rangle + \big\langle (D_t)^*(f_2^n)_{s-1}, (\zeta^{n+1})_{s-1}\big\rangle\\ &+ \frac{d}{dt}\big\langle (f_2^n)_{s-1}, (\zeta^{n+1})_{s-1}\big\rangle,\end{aligned}$$

from which, Lemma 2.19, Lemma 11.16 and Lemma 11.1, we deduce that

$$\begin{aligned}\big\langle (f_2^n)_{s-1},(D_t\zeta^{n+1})_{s-1}\big\rangle =&\frac{d}{dt}\big\langle (f_2^n)_{s-1},(\zeta^{n+1})_{s-1}\big\rangle\\&+\mathcal{A}_2\big(\|f_2^n\|_{H^{s-\frac32}}+\|D_tf_2^n\|_{H^{s-\frac32}}\big)\|\zeta^{n+1}\|_{H^{s-\frac12}}\\=&\frac{d}{dt}\big\langle (f_2^n)_{s-1},(\zeta^{n+1})_{s-1}\big\rangle+\mathcal{A}_5\mathcal{E}^{n+1}(t).\end{aligned}$$

In a similar way, we deduce from Lemma 11.18 and Lemma 11.17 that

$$\big\langle (D_tf_\omega^n)_{s-1},(D_t\zeta^{n+1})_{s-1}\big\rangle=\frac{d}{dt}\big\langle (D_t^uf_\omega^n)_{s-1},(\zeta^{n+1})_{s-1}\big\rangle+\mathcal{A}_6\mathcal{E}^{n+1}(t),$$

$$\big\langle (D_tf_4^n)_{s-1},(D_t\zeta^{n+1})_{s-1}\big\rangle=\frac{d}{dt}\big\langle (D_t^uf_4^n)_{s-1},(\zeta^{n+1})_{s-1}\big\rangle+\mathcal{A}_6\mathcal{E}^{n+1}(t).$$

We write

$$\begin{aligned}\big\langle (D_tf_3^n)_{s-1},(D_t\zeta^{n+1})_{s-1}\big\rangle=&\big\langle [\langle D\rangle^{s-1},D_t]f_3^n,(D_t\zeta^{n+1})_{s-1})\big\rangle\\&+\big\langle (f_3^n)_{s-1},(D_t)^*(D_t\zeta^{n+1})_{s-1}\big\rangle\\&+\frac{d}{dt}\big\langle (f_3^n)_{s-1},(D_t\zeta^{n+1})_{s-1}\big\rangle,\end{aligned}$$

which along with Lemma 2.19 and Lemma 11.17 gives

$$\begin{aligned}\big\langle (D_tf_3^n)_{s-1},(D_t\zeta^{n+1})_{s-1}\big\rangle =&\frac{d}{dt}\big\langle (f_3^n)_{s-1},(D_t\zeta^{n+1})_{s-1}\big\rangle\\&+\mathcal{A}_2\|f_3^n\|_{H^{s-\frac12}}\big(\|D_t\zeta^{n+1}\|_{H^{s-1}}+\|D_t^2\zeta^{n+1}\|_{H^{s-\frac32}}\big)\\=&\frac{d}{dt}\big\langle (f_3^n)_{s-1},(D_t\zeta^{n+1})_{s-1}\big\rangle+\mathcal{A}_5\mathcal{E}^{n+1}(t)+\mathcal{A}_5.\end{aligned}$$

Putting the above estimates together, we obtain

$$\begin{aligned}\frac12\frac{d}{dt}\big(\|D_t\zeta^{n+1}\|_{H^{s-1}}^2+\|T_{\sqrt{a^n\lambda^n}}\zeta^{n+1}\|_{H^{s-1}}^2\big)\le\mathcal{A}_5\mathcal{E}^{n+1}(t)^2+\mathcal{A}_6\\+\frac{d}{dt}\big\langle (f_2^n)_{s-1}+(f_3^n)_{s-1}+(D_t^uf_4^n)_{s-1}+(D_t^uf_\omega^n)_{s-1},(\zeta^{n+1})_{s-1}\big\rangle.\end{aligned}$$

Integrating in t and using Lemma 11.16-Lemma 11.18, we deduce that

$$\begin{aligned}&\|D_t\zeta^{n+1}\|_{H^{s-1}}^2+\|T_{\sqrt{a^n\lambda^n}}\zeta^{n+1}\|_{H^{s-1}}^2\le\mathcal{P}(\mathcal{E}_0)+\mathcal{A}_6\int_0^t(1+\mathcal{E}^{n+1}(t')^2)dt'\\&\qquad+\mathcal{P}(E_1,E_2,E_4)+\mathcal{A}_1\|\eta^n\|_{H^s}^2+\epsilon\|\zeta^{n+1}\|_{H^{s-\frac12}}^2\\&\le\mathcal{P}(\mathcal{E}_0)+\mathcal{P}(E_1,E_2,E_4)+\mathcal{A}_6(1+\mathcal{P}_0)^2t+\frac{1}{64}\mathcal{P}_0^2+\frac14\|T_{\sqrt{a^n\lambda^n}}\zeta^{n+1}\|_{H^{s-1}}^2.\end{aligned}$$

This implies that there exists $T_4>0$ depending on $\mathcal{P}_0,\mathcal{P}_1$ so that for $t\in[0,\min(T,T_n)]$ with $T=\min(T_1,T_2,T_3,T_4)$,

$$\mathcal{E}^{n+1}(t)\le\mathcal{P}(\mathcal{E}_0)+\mathcal{P}(E_1,E_2,E_4)+\mathcal{P}_1+\frac14\mathcal{P}_0.$$

Since E_1,E_2,E_3 are independent of $\mathcal{P}_0$, we can take $\mathcal{P}_0=4\big(\mathcal{P}(\mathcal{E}_0)+\mathcal{P}(E_1,E_2,E_4)+\mathcal{P}_1\big)$. Thus, we deduce that for $t\in[0,\min(T,T_n)]$,

$$\mathcal{E}^{n+1}(t)\le\frac{\mathcal{P}_0}{2}.$$

This completes the proof of uniform estimates. □

CHAPTER 12

Cauchy sequence and the limit system

This chapter is devoted to showing that the approximate sequence constructed in the last chapter is a Cauchy sequence.

12.1. Set-up

According to the uniform estimates of the approximate sequence, we may assume that

$$\sum_{k=0}^{2}\Big(\|\partial_t^k(V^n,B^n,V_b^n,V_1^n,B_1^n,V_{b,1}^n)\|_{H^{s-\frac12-k}}+\|\partial_t^k\eta^n\|_{H^{s+\frac12-k}}+\|\partial_t^k\eta_1^n\|_{H^{s-\frac12-k}}$$
$$+\|\partial_t^k\widetilde{\omega^n}\|_{H^{s-1-k}(\overline{\mathcal{S}})}\Big)\le\mathcal{P}_0.$$

Here and in what follows we denote by $\mathcal{P}_0$ a constant depending only on $\|v_0\|_{H^s(\Omega_0)}$, $\|\eta_0\|_{H^{s+\frac12}}$ and h_0,c_0, which may change from line to line. For a function $f(x,y)$ defined on $\{(x,y):-1<y<\eta(x)\}$, we denote $\widetilde{f}(x,z)\triangleq f(x,\rho_\delta(x,z))$, where $\rho_\delta(x,z)=z+(1+z)e^{\delta z|D|}\eta$.

We introduce

$$\delta_{\mathcal{E}^n}(t)\overset{\text{def}}{=}\delta_{\mathcal{E}_1^n}(t)+\delta_{\mathcal{E}_2^n}(t),$$

where $\delta_{\mathcal{E}_1^n}(t)$ and $\delta_{\mathcal{E}_2^n}(t)$ are given by

$$\begin{aligned}\delta_{\mathcal{E}_1^n}(t)=&\|D_t\delta_{\zeta^n}(t)\|_{H^{s-2}}+\|T_{\sqrt{a^n\lambda^n}}\delta_{\zeta^n}(t)\|_{H^{s-2}},\\ \delta_{\mathcal{E}_2^n}(t)=&\|\big(\delta_{V^n},\delta_{B^n},\delta_{V_b^n},\delta_{V_1^n},\delta_{B_1^n},\delta_{V_{b,1}^n},\delta_{\eta_1^n},\delta_{\eta^n}\big)(t)\|_{H^{s-\frac32}}+\|\delta_{\widetilde{\omega^n}}(t)\|_{H^{s-2}(\overline{\mathcal{S}})}.\end{aligned}$$

Here $D_t=\partial_t+T_{V^{n+1}}\cdot\nabla$ and we denote $\delta_{f^n}\triangleq f^{n+1}-f^n$.

Throughout this section, we denote by $L_i^n(i=1,2)$ some nonlinear terms, which satisfy

$$\|L_i^n(t)\|_{H^{s-\frac12-i}}\le\mathcal{P}_0\big(\delta_{\mathcal{E}^n}(t)+\delta_{\mathcal{E}^{n-1}}(t)\big).$$

Using Lemma 2.10 and Proposition 2.6, it is easy to find that

$$\begin{cases}D_t\big(\delta_{V^n},\delta_{B^n}\big)=L_1^n+L(\delta_{a^{n-1}}),\\ (\partial_t+V_b^n\cdot\nabla)\delta_{V_b^n}=-\delta_{\nabla P^{n-1}|_{y=-1}},\\ D_t\delta_{\zeta^n}=L_2^n+L_3^n+\delta_{F_2},\\ (\partial_t+T_{V^{n+1}}\cdot\nabla)(\delta_{V_1^n},\delta_{B_1^n})=L_1^n+L(\delta_{a^{n-1}}),\\ (\partial_t+V_b^{n+1}\cdot\nabla)\delta_{V_{b,1}^n}=-\delta_{\nabla P^{n-1}|_{y=-1}},\\ (\partial_t+V^{n+1}\cdot\nabla)\delta_{\eta_1^n}=L_1^n,\end{cases}\tag{12.1}$$

where $L(\delta_{a^{n-1}})$ is a nonlinear term satisfying

$$\|L(\delta_{a^{n-1}})\|_{H^{s-\frac{3}{2}}} \leq C\|\delta_{a^{n-1}}\|_{H^{s-\frac{3}{2}}},$$

and L_3^n is given by

$$L_3^n = (R(\eta_1^n) - R(\eta_1^{n-1}))V_1^n + \zeta^n(R(\eta_1^n) - R(\eta_1^{n-1}))B_1^n + (R_\omega^n - R_\omega^{n-1}).$$

12.2. Elliptic estimates with a parameter

In order to compare the solution of the elliptic equations in two different domains and with different boundary values, we consider the following elliptic equation in a domain $\Omega_\tau(t) = \{(x,y) : -1 < y < \eta(\tau,t,x), x \in \mathbf{R}^d\}$ with a parameter $\tau \in [0,1]$:

$$\begin{cases} \Delta_{x,y} v(\tau,t,x,y) = F(\tau,t,x) \quad \text{in} \quad \Omega_\tau, \\ v|_{y=\eta(\tau,t,x)} = V(\tau,t,x), \quad v|_{y=-1} = V^b(\tau,t,x). \end{cases} \tag{12.2}$$

In the sequel, we denote $f_\tau \triangleq \partial_\tau f(\tau,t,x,y)$. Then v_τ satisfies

$$\begin{cases} \Delta_{x,y} v_\tau = F_\tau \quad \text{in} \quad \Omega_\tau, \\ v_\tau|_{y=\eta(\tau,t,x)} = V_\tau - \partial_y v \eta_\tau, \quad v_\tau|_{y=-1} = V_\tau^b. \end{cases}$$

If (η, V, V^b) and F satisfy

$$\begin{aligned} &\|\eta\|_{H^{s-\frac{1}{2}}} + \|V\|_{H^{s-\frac{1}{2}}} + \|V_\tau^b\|_{H^{s-\frac{1}{2}}} + \|\widetilde{F_\tau}\|_{H^{s-\frac{1}{2}}(\mathcal{S})} \leq C, \\ &\|\eta_\tau\|_{H^{s-\frac{3}{2}}} + \|V_\tau\|_{H^{s-\frac{3}{2}}} + \|V_\tau^b\|_{H^{s-\frac{3}{2}}} + \|\widetilde{F_\tau}\|_{H^{s-\frac{3}{2}}(\mathcal{S})} \leq C, \\ &\eta(\tau,t,x) + 1 \geq \frac{h_0}{2} \quad \text{for} \quad (\tau,t,x) \in [0,1] \times [0,T] \times \mathbf{R}^d, \end{aligned}$$

then we infer from Proposition 4.9 and Proposition 4.10 that for any $\sigma \in [-\frac{1}{2}, s-\frac{5}{2}]$,

$$\|\nabla_{x,z}\widetilde{v_\tau}\|_{X^\sigma([-\frac{1}{2},0])} \leq C\big(\|\widetilde{F_\tau}\|_{L^2(H^{\sigma-\frac{1}{2}}(\mathcal{S}))} + \|(V_\tau, \eta_\tau)\|_{H^{\sigma+1}} + \|V_\tau^b\|_{H^{\frac{1}{2}}}\big), \tag{12.3}$$

$$\|\nabla_{x,z}\widetilde{v_\tau}\|_{H^{\sigma+\frac{1}{2}}(\mathcal{S})} \leq C\big(\|\widetilde{F_\tau}\|_{H^{\sigma-\frac{1}{2}}(\mathcal{S})} + \|(V_\tau, V_\tau^b, \eta_\tau)\|_{H^{\sigma+1}}\big). \tag{12.4}$$

Now we take in the elliptic equation (12.2):

$$\begin{aligned} &\eta = \tau\eta_1^{n+1} + (1-\tau)\eta_1^n, \\ &V = \tau(V^{n+1}, B^{n+1}) + (1-\tau)(V^n, B^n), \\ &V^b = \tau(V_b^{n+1}, 0) + (1-\tau)(V_b^n, 0), \\ &F = \nabla_{x,y}\cdot\big(\tau\widetilde{\omega^{n+1}} \circ \Phi^{-1} + (1-\tau)\widetilde{\omega^n} \circ \Phi^{-1}\big), \end{aligned}$$

where $\Phi(\tau,x,z) = (x, \rho_\delta(\tau,x,z))$ with $\rho_\delta(\tau,x,z) = z + (1+z)e^{\delta z|D|}\eta(\tau,t,\cdot)$. Notice that

$$\delta_{\widetilde{v^n}} = \widetilde{v^{n+1}} - \widetilde{v^n} = \int_0^1 \widetilde{v_\tau}\,d\tau, \quad \eta_\tau = \delta_{\eta^n}, \quad V_\tau = \delta_{(V^n,B^n)}, \quad V_\tau^b = \delta_{(V_b^n,0)},$$

$$\|\widetilde{F_\tau}\|_{H^{s-3}(\mathcal{S})} \leq \mathcal{P}_0\delta_{\mathcal{E}_2^n}(t).$$

Then we can deduce from (12.4) that

$$\|\nabla_{x,z}\delta_{\widetilde{v^n}}\|_{H^{s-2}(\mathcal{S})} \leq \mathcal{P}_0\delta_{\mathcal{E}_2^n}(t). \tag{12.5}$$

In a similar way, we can deduce that

$$\|\nabla_{x,z}\delta_{\widetilde{v_1^n}}\|_{H^{s-2}(\mathcal{S})} \leq \mathcal{P}_0\delta_{\mathcal{E}_2^n}(t). \tag{12.6}$$

Let $\omega(\tau,t,x) = \tau\widetilde{\omega^{n+1}}\circ\Phi^{-1} + (1-\tau)\widetilde{\omega^n}\circ\Phi^{-1}$ and

$$\widetilde{\mathcal{D}}_t = \partial_t + (\widetilde{v_2^{n+1}}\circ\Phi^{-1} - \partial_t\Phi^{-1})\cdot\mathcal{A}\nabla_{x,y}$$

with $\mathcal{A} = \left(\partial_i\Phi_j^{-1}\right)^{-1}$. Then ω_τ satisfies

$$\|(\widetilde{\mathcal{D}}_t\omega_\tau)\circ\Phi\|_{H^{s-2}(\overline{\mathcal{S}})} \le \mathcal{P}_0\delta_{\mathcal{E}_2^n}(t). \tag{12.7}$$

This in turn implies that

$$\|\delta_{\widetilde{\mathcal{D}_t v_1^n}}\|_{H^{s-1}(\overline{\mathcal{S}})} + \|(\widetilde{\mathcal{D}}_t^2\omega_\tau)\circ\Phi\|_{H^{s-2}(\overline{\mathcal{S}})} \le \mathcal{P}_0\delta_{\mathcal{E}^n}(t). \tag{12.8}$$

Similarly, there holds for $\delta_{v_\omega^{n-1}}$,

$$\|\delta_{\widetilde{v_\omega^{n-1}}}\|_{H^{s-1}(\overline{\mathcal{S}})} + \|\delta_{\widetilde{\mathcal{D}_t v_\omega^{n-1}}}\|_{X^{s-\frac{3}{2}}([-\frac{1}{2},0])} \le \mathcal{P}_0\delta_{\mathcal{E}_2^{n-1}}(t), \tag{12.9}$$

$$\|\delta_{\widetilde{\mathcal{D}_t^2 v_\omega^{n-1}}}\|_{X^{s-\frac{3}{2}}([-\frac{1}{2},0])} \le \mathcal{P}_0\delta_{\mathcal{E}^{n-1}}(t). \tag{12.10}$$

To compare the pressure, we consider

$$\begin{cases} -\Delta_{x,y}P(\tau,t,x,y) = F(\tau,t,x) \quad \text{in} \quad \Omega_\tau, \\ P|_{y=\eta(\tau,t,x)} = 0, \quad \partial_y P|_{y=-1} = -1. \end{cases}$$

where $\eta(\tau,t,x) = \tau\eta^n + (1-\tau)\eta^{n-1}$ and

$$F = \left(\left(\partial_i({v_1}^j)^n\partial_j({v_1}^i)^n\right)\circ\Phi_1^n - \left(\partial_i({v_1}^j)^{n-1}\partial_j({v_1}^i)^{n-1}\right)\circ\Phi_1^{n-1}\right)\circ\Phi^{-1}.$$

Then P_τ satisfies

$$\begin{cases} -\Delta_{x,y}P_\tau = F_\tau \quad \text{in} \quad \Omega_\tau, \\ P_\tau|_{y=\eta(\tau,t,x)} = -\partial_y P\eta_\tau, \quad \partial_y P_\tau|_{y=-1} = 0. \end{cases}$$

Thus, we can deduce by a similar proof of Lemma 11.6 that

$$\|\delta_{a^{n-1}}\|_{H^{s-\frac{5}{2}}} + \|\delta_{\nabla P^{n-1}|_{y=-1}}\|_{H^{s-\frac{5}{2}}} \le \mathcal{P}_0\delta_{\mathcal{E}_2^{n-1}}(t), \tag{12.11}$$

$$\|\delta_{a^{n-1}}\|_{H^{s-\frac{3}{2}}} + \|\delta_{\nabla P^{n-1}|_{y=-1}}\|_{H^{s-\frac{3}{2}}} \le \mathcal{P}_0\delta_{\mathcal{E}^{n-1}}(t), \tag{12.12}$$

$$\|\delta_{\nabla\partial_t P^{n-1}|_{y=-1}}\|_{H^{s-\frac{5}{2}}} + \|\delta_{\partial_t a^{n-1}}\|_{H^{s-\frac{5}{2}}} \le \mathcal{P}_0\delta_{\mathcal{E}^{n-1}}(t). \tag{12.13}$$

To compare the remainder of (DN) operators, we take in (12.2):

$$\eta = \tau\eta_1^n + (1-\tau)\eta_1^{n-1}, \quad V = (V_1^n, B_1^n), \quad V^b = 0, \quad F = 0.$$

Then v_τ satisfies

$$\begin{cases} \Delta_{x,y}v_\tau = 0 \quad \text{in} \quad \Omega_\tau, \\ v_\tau|_{y=\eta(\tau,t,x)} = -\partial_y v\eta_\tau, \quad v_\tau|_{y=-1} = 0. \end{cases}$$

Notice that

$$\|D_t^u\eta_\tau\|_{H^{s-\frac{3}{2}}} \le \mathcal{P}_0\delta_{\mathcal{E}_2^{n-1}}(t), \quad \|(D_t^u)^2\eta_\tau\|_{H^{s-\frac{5}{2}}} \le \mathcal{P}_0\delta_{\mathcal{E}^{n-1}}(t).$$

Here $D_t^u = \partial_t + V^n\cdot\nabla$. Thus, we can deduce from a similar proof of Proposition 11.7 that

$$\|(R(\eta_1^n) - R(\eta_1^{n-1}))(V_1^n, B_1^n)\|_{H^{s-\frac{5}{2}}} \le \mathcal{P}_0\delta_{\mathcal{E}_2^{n-1}}(t), \tag{12.14}$$

$$\|D_t^u(R(\eta_1^n) - R(\eta_1^{n-1}))(V_1^n, B_1^n)\|_{H^{s-\frac{5}{2}}} \le \mathcal{P}_0\delta_{\mathcal{E}_2^{n-1}}(t), \tag{12.15}$$

$$\|(D_t^u)^2(R(\eta_1^n) - R(\eta_1^{n-1}))(V_1^n, B_1^n)\|_{H^{s-\frac{5}{2}}} \le \mathcal{P}_0\delta_{\mathcal{E}^{n-1}}(t). \tag{12.16}$$

12.3. Energy estimates

Following the proofs in section 11.7 and using the estimates in section 12.4, we can deduce that

$$\|\delta_{f_1^{n-1}}\|_{H^{s-2}} + \|\delta_{f_3^{n-1}}\|_{H^{s-\frac{3}{2}}} \leq \mathcal{P}_0\big(\delta_{\mathcal{E}^n}(t) + \delta_{\mathcal{E}^{n-1}}(t)\big),$$
$$\|\delta_{f_3^{n-1}}\|_{H^{s-2}} \leq \mathcal{P}_0\big(\|\delta_{\eta^{n-1}}\|_{H^{s-1}} + \delta_{\mathcal{E}_2^{n-1}}(t)\big),$$
$$\|\delta_{f_2^{n-1}}\|_{H^{s-\frac{5}{2}}} + \|D_t^u\delta_{f_4^{n-1}}\|_{H^{s-\frac{5}{2}}} + \|D_t^u\delta_{f_\omega^{n-1}}\|_{H^{s-\frac{5}{2}}} \leq \mathcal{P}_0\big(\delta_{\mathcal{E}_2^{n-1}}(t) + \delta_{\mathcal{E}_2^n}(t)\big),$$
$$\|D_t\delta_{f_2^{n-1}}\|_{H^{s-\frac{5}{2}}} + \|(D_t^u)^2\delta_{f_4^{n-1}}\|_{H^{s-\frac{5}{2}}} + \|(D_t^u)^2\delta_{f_\omega^{n-1}}\|_{H^{s-\frac{5}{2}}} \leq \mathcal{P}_0\big(\delta_{\mathcal{E}^n}(t) + \delta_{\mathcal{E}^{n-1}}(t)\big).$$

By (12.14) and (12.9), we have

$$\|L_3^n\|_{H^{s-\frac{5}{2}}} \leq \mathcal{P}_0\delta_{\mathcal{E}_2^{n-1}}(t).$$

We also have

$$D_t^2\delta_{\zeta^n} + T_{a^n\lambda^n}\delta_{\zeta^n} = L_2^n + L_4^n, \tag{12.17}$$

where L_4^n is given by

$$L_4^n = (f_2^n + D_tf_3^n + D_tf_4^n + D_tf_\omega^n) - (f_2^{n-1} + D_tf_3^{n-1} + D_tf_4^{n-1} + D_tf_\omega^{n-1}).$$

For $\delta_{\widetilde{\omega^n}}$, we have

$$\|(\partial_t + \overline{v}^{n+1}\cdot\nabla_{x,z})\delta_{\widetilde{\omega^n}}\|_{H^{s-2}(\overline{\mathcal{S}})} \leq \mathcal{P}_0\delta_{\mathcal{E}_2^{n-1}}(t).$$

Making the energy estimates for the system (12.1), we obtain

$$\delta_{\mathcal{E}_2^n}(t)^2 \leq \mathcal{P}_0\int_0^t \big(\delta_{\mathcal{E}^n}(t') + \delta_{\mathcal{E}^{n-1}}(t')\big)^2 dt'.$$

While, making the energy estimates for the system (12.17), we obtain

$$\begin{aligned}\delta_{\mathcal{E}_1^n}(t)^2 \leq & \mathcal{P}_0\int_0^t \big(\delta_{\mathcal{E}^n}(t') + \delta_{\mathcal{E}^{n-1}}(t')\big)^2 dt' + \frac{1}{2}\delta_{\mathcal{E}^{n-1}}(t)^2 \\ & + \mathcal{P}_0\big(\delta_{\mathcal{E}_2^n}(t) + \delta_{\mathcal{E}_2^{n-1}}(t)\big)^2.\end{aligned}$$

Then an induction argument ensures that there exists $T > 0$ depending only on $\mathcal{P}_0$ so that for $t \in [0, T]$,

$$\delta_{\mathcal{E}^n}(t) \longrightarrow 0 \quad \text{as} \quad n \longrightarrow +\infty.$$

This shows that the approximate sequence is a Cauchy sequence.

12.4. The limit system

Let $\big(V, B, V_b, V_1, B_1, V_{b,1}, \zeta, \eta_1, \eta, \omega, v, v_1, P\big)$ be the limit of the Cauchy sequence

$$\big(V^n, B^n, V_b^n, V_1^n, B_1^n, V_{b,1}^n, \zeta^n, \eta^n, \eta_1^n, \omega^n, v^n, v_1^n, P^n\big).$$

Taking the limit for the approximate system (10.13)–(10.11), we obtain the following limit system: The boundary velocity (V, B, V_b) and ζ satisfy

$$
\begin{cases} D_t V = T_\zeta a + T_a \zeta + R(\zeta, a) + (T_{V_1} - V_1)\cdot\nabla V, \\ D_t B = a - 1 + (T_{V_1} - V_1)\cdot\nabla B, \\ (\partial_t + V_b\cdot\nabla)V_b = -\nabla P|_{y=-1}, \\ D_t\zeta = T_\lambda(V + T_\zeta B) + (T_{V_1} - V_1)\cdot\zeta + [T_\zeta, T_\lambda]B_1 \\ \qquad + (\zeta - T_\zeta)T_\lambda B + R(\eta_1)V_1 + \zeta R(\eta_1)B_1 + R_\omega, \end{cases} \tag{12.18}
$$

where $D_t = \partial_t + T_V\cdot\nabla, a = -\partial_y P|_{y=\eta}, \lambda = \lambda(\eta_1)$, and

$$
\begin{aligned}
(R_\omega)^i =& \big(\partial_y(v_\omega)^i - \partial_{x_j}(v_\omega)^i\cdot\partial_{x_j}\eta_1\big) + \partial_{x_i}\eta_1\big(\partial_y(v_\omega)^{d+1} - \partial_{x_j}\eta_1\partial_{x_j}(v_\omega)^{d+1}\big) \\
&+ \big(\omega_{i,d+1} - \partial_{x_j}\eta_1\omega_{ij} + \partial_{x_i}\eta_1\partial_{x_j}\eta_1\omega_{j,d+1}\big)\big|_{y=\eta_1}
\end{aligned}
$$

with v_ω given by

$$
\begin{cases} -\Delta_{x,y}v_\omega = \nabla_{x,y}\times\omega \quad \text{in} \quad \Omega_t = \big\{(x,y) : -1 < y < \eta(t,x)\big\}, \\ v_\omega|_{y=\eta_1} = 0, \quad v_\omega|_{y=-1} = (V_b, 0), \end{cases}
$$

and $(V_1, B_1, V_{b,1})$ satisfies

$$
\begin{cases} D_t(V_1, B_1) = D_t(V, B), \\ (\partial_t + V_b\cdot\nabla)V_{b,1} = -\nabla P|_{y=-1}. \end{cases} \tag{12.19}
$$

The free surface (η, η_1) satisfies

$$
-\Delta\eta + \eta = -\operatorname{div}\zeta + \eta_1, \tag{12.20}
$$

$$
(\partial_t + V\cdot\nabla)\eta_1 = B_1. \tag{12.21}
$$

The vorticity ω satisfies

$$
\partial_t\omega + \big(v^h\cdot\nabla + v_1^{d+1}\partial_y\big)\omega = \omega\cdot\nabla_{x,y}v_1, \tag{12.22}
$$

where the velocity (v, v_1) is given by

$$
\begin{cases} -\Delta_{x,y}v = \nabla_{x,y}\times\omega \quad \text{in} \quad \Omega_t, \\ v\,|_{y=\eta_1} = (V, B), \quad v|_{y=-1} = (V_b, 0), \end{cases} \tag{12.23}
$$

and

$$
\begin{cases} -\Delta_{x,y}v_1 = \nabla_{x,y}\times\omega \quad \text{in} \quad \Omega_t, \\ v_1\,|_{y=\eta_1} = (V_1, B_1), \quad v_1|_{\Gamma_b} = (V_{b,1}, 0). \end{cases} \tag{12.24}
$$

Let $\widetilde{\Omega}_t = \big\{(x,y) : y = \eta(t,x)\big\}$. The pressure P satisfies

$$
\begin{cases} -\Delta_{x,y}P = \big(\partial_i({v_1}^j)\partial_j({v_1}^i)\big)\circ\Phi_1\circ(\Phi)^{-1} \quad \text{in} \quad \widetilde{\Omega}_t, \\ P\,|_{y=\eta} = 0, \quad \partial_y P\,|_{y=-1} = 0, \end{cases} \tag{12.25}
$$

where $\Phi(x,z) = (x, \rho_{\delta,\eta}(x,z))$ and $\Phi_1(x,z) = (x, \rho_{\delta,\eta_1}(x,z))$ with $\rho_{\delta,\eta}(x,z) = z + (1+z)e^{\delta z|D|}\eta$.

CHAPTER 13

From the limit system to the Euler equations

The goal of this chapter is to show that the limit system (12.18)–(12.25) is equivalent to the Euler equations (1.1)–(1.5).

First of all, it follows from (12.19) and the third equation of (10.13) that

$$(V, B, V_b) = (V_1, B_1, V_{b,1}).$$

Hence, $v = v_1$ since they satisfy the same elliptic equation with the same boundary conditions. Thus, we deduce from (12.18) that

$$\begin{cases} (\partial_t + V \cdot \nabla)V = a\zeta, \quad a = \nabla P|_{y=\eta}, \\ (\partial_t + V \cdot \nabla)B = a - 1, \\ (\partial_t + V_b \cdot \nabla)V_b = -\nabla P|_{y=-1}, \\ (\partial_t + V \cdot \nabla)\zeta = G(\eta_1)V + \zeta G(\eta_1)B + R_\omega. \end{cases} \tag{13.1}$$

For the free surface, we have

$$(\partial_t + V \cdot \nabla)\eta_1 = B. \tag{13.2}$$

The vorticity ω satisfies

$$\begin{cases} \partial_t\omega + v \cdot \nabla_{x,y}\omega = \omega \cdot \nabla_{x,y} v \quad \text{in} \quad \Omega_t, \\ -\Delta_{x,y} v = \nabla_{x,y} \times \omega \quad \text{in} \quad \Omega_t, \\ v\ |_{y=\eta_1} = (V, B), \quad v\ |_{y=-1} = (V_b, 0). \end{cases} \tag{13.3}$$

It remains to show that

$$\eta = \eta_1, \quad \omega = \nabla_{x,y} \times v, \quad \mathrm{div} v = 0. \tag{13.4}$$

For this end, we introduce

$$G = v_t + v \cdot \nabla_{x,y} v + (\nabla_{x,y}(P + gy)) \circ \overline{\Phi},$$
$$\delta_\omega = \omega - \overline{\omega}, \quad \delta_d = \mathrm{div} v, \quad \delta_\zeta = \zeta - \nabla\eta_1.$$

where $\overline{\omega} = \nabla_{x,y} \times v$ and $\overline{\Phi} = \Phi \circ (\Phi_1)^{-1}$.

In what follows, we denote by $L(\cdot)$ a linear function, which may be different from line to line.

LEMMA 13.1. *It holds that*

$$\overline{\Phi}(x, y) = (x, h(x, y)),$$

where $h(x, y)$ satisfies

$$\|\nabla_{x,y}(h - y)\|_{H^1(\Omega_t)} \le C\|\eta - \eta_1\|_{H^{\frac{3}{2}}}.$$

PROOF. Let $\Phi_1^{-1}(x,y)=(x,z(x,y))$, i.e.,

$$y=z(x,y)+(1+z(x,y))e^{\delta z(x,y)|D|}\eta_1.$$

Hence, we have

$$\overline{\Phi}(x,y)=\big(x,z(x,y)+(1+z(x,y))e^{\delta z(x,y)|D|}\eta\big)\triangleq(x,h(x,y)).$$

Using the fact that

$$\begin{aligned}&\nabla z(x,y)(1+e^{\delta z(x,y)|D|}\eta_1)\\&\qquad+(1+z(x,y))(\nabla z(x,y)e^{\delta z(x,y)|D|}|D|\eta_1+e^{\delta z(x,y)|D|}\nabla\eta_1)=0,\\&\partial_y z(x,y)(1+e^{\delta z(x,y)|D|}\eta_1)+(1+z(x,y))\partial_y z(x,y)e^{\delta z(x,y)|D|}|D|\eta_1=1,\end{aligned}$$

we deduce that

$$\nabla_{x,y}(h(x,y)-y)=L\big(e^{\delta z(x,y)|D|}(\eta-\eta_1),e^{\delta z(x,y)|D|}\nabla(\eta-\eta_1)\big),$$

which implies that

$$\begin{aligned}\|\nabla_{x,y}(h-y)\|_{H^1(\Omega_t)}\leq&C\|e^{\delta z(x,y)|D|}(\eta-\eta_1)\|_{H^1(\Omega_t)}+C\|e^{\delta z(x,y)|D|}\nabla(\eta-\eta_1)\|_{H^1(\Omega_t)}\\\leq&C\|e^{\delta z|D|}(\eta-\eta_1)\|_{L^2_z(-1,0;H^2)}\leq C\|\eta-\eta_1\|_{H^{\frac32}}.\end{aligned}$$

The proof is finished. □

Let us first derive the equation of G.

LEMMA 13.2. *It holds that*

$$\begin{cases}\Delta_{x,y}G=L\big(\delta_d,\nabla_{x,y}\delta_d,\nabla^2_{x,y}\delta_d,\delta_\omega,\nabla_{x,y}\delta_\omega,\nabla^2_{x,y}\delta_\omega,\nabla_{x,y}(h-y),\nabla^2_{x,y}h\big),\\ G|_{y=\eta_1}=0,\quad G|_{y=-1}=0.\end{cases}$$

In particular, we have

$$\|G\|_{H^2(\Omega_t)}\leq C\big(\|(\delta_d,\delta_\omega)\|_{L^2(\Omega_t)}+\|\nabla_{x,y}(h-y)\|_{H^1(\omega_t)}\big).$$

PROOF. Thanks to (13.3), we have

$$\Delta_{x,y}v_t=-\nabla_{x,y}\times\omega_t=\nabla_{x,y}\times\big(v\cdot\nabla_{x,y}\omega-\omega\cdot\nabla_{x,y}v\big).$$

A direct calculation gives

$$\begin{aligned}\Delta_{x,y}\big(v\cdot\nabla_{x,y}v\big)=&-\nabla_{x,y}\times\big(\nabla_{x,y}\times(v\cdot\nabla v)\big)+\nabla_{x,y}\mathrm{div}(v\cdot\nabla_{x,y}v)\\=&-\nabla_{x,y}\times\big(v\cdot\nabla_{x,y}\overline{\omega}-\overline{\omega}\cdot\nabla_{x,y}v+\overline{\omega}\delta_d\big)\\&+\nabla_{x,y}(\partial_jv^i\partial_iv^j)+\nabla_{x,y}(v\cdot\nabla_{x,y}\delta_d),\end{aligned}$$

Using (12.25) and Lemma 13.1, we obtain

$$\Delta_{x,y}\big(\nabla_{x,y}P\circ\overline{\Phi}\big)=-\nabla_{x,y}(\partial_jv^i\partial_iv^j)+L\big(\nabla_{x,y}(h-y),\nabla^2_{x,y}h\big).$$

The first equation of the lemma follows by summing up the above equations. The boundary condition of G follows from (13.1). □

Now we have

$$v_t+v\cdot\nabla_{x,y}v+(\nabla_{x,y}(P+gy))\circ\overline{\Phi}=G.\tag{13.5}$$

Take the divergence to get

$$\partial_t\delta_d+v\cdot\nabla_{x,y}\delta_d=\mathrm{div}_{x,y}G+L(\nabla_{x,y}(h-y)),$$

which along with Lemma 13.1 and Lemma 13.2 implies that

$$\begin{aligned}\frac{d}{dt}\|\delta_d\|^2_{H^1(\Omega_t)} \leq& C\big(\|\delta_d\|^2_{H^1(\Omega_t)} + \|G\|_{H^2(\Omega_t)} + \|\nabla_{x,y}(h-y)\|_{H^1(\Omega_t)}\big)\\ \leq& C\big(\|(\delta_d,\delta_\omega)\|^2_{H^1(\Omega_t)} + \|\eta-\eta_1\|_{H^{\frac{3}{2}}}\big).\end{aligned} \tag{13.6}$$

Taking the curl to (13.5), we obtain

$$\partial_t\overline{\omega} + v\cdot\nabla_{x,y}\overline{\omega} = \overline{\omega}\cdot\nabla_{x,y}v + \nabla_{x,y}\times G + L(\nabla_{x,y}(h-y)),$$

from which and (13.3), we infer

$$\partial_t\delta_\omega + v\cdot\nabla_{x,y}\delta_\omega = \delta_\omega\cdot\nabla_{x,y}v + \nabla_{x,y}\times G + L(\nabla_{x,y}(h-y)).$$

Then similar to (13.6), we have

$$\frac{d}{dt}\|\delta_\omega\|^2_{H^1(\Omega_t)} \leq C\big(\|(\delta_d,\delta_\omega)\|^2_{H^1(\Omega_t)} + \|\eta-\eta_1\|_{H^{\frac{3}{2}}}\big). \tag{13.7}$$

Let $\zeta_1 = \nabla\eta_1$. By a similar derivation of (6.10), we get

$$\partial_t\zeta_1 + V\cdot\nabla\zeta_1 = G(\eta_1)V + \zeta_1 G(\eta_1)B + R_{\overline{\omega}},$$

where $R_{\overline{\omega}}$ is given by

$$\begin{aligned}(R_{\overline{\omega}})^i =& \big(\partial_y(v_\omega)^i - \partial_{x_j}(v_\omega)^i\cdot\partial_{x_j}\eta_1\big) + \partial_{x_i}\eta_1\big(\partial_y(v_\omega)^{d+1} - \partial_{x_j}\eta_1\partial_{x_j}(v_\omega)^{d+1}\big)\\ &+ \big(\overline{\omega}_{i,d+1} - \partial_{x_j}\eta_1\overline{\omega}_{ij} + \partial_{x_i}\eta_1\partial_{x_j}\eta_1\overline{\omega}_{j,d+1}\big)\big|_{y=\eta_1}.\end{aligned}$$

Hence, δ_ζ satisfies

$$\partial_t\delta_\zeta + V\cdot\nabla\delta_\zeta = \delta_\zeta G(\eta_1)B + R_\omega - R_{\overline{\omega}}.$$

This ensures that

$$\frac{d}{dt}\|\delta_\zeta\|^2_{H^{\frac{1}{2}}} \leq C\big(\|\delta_\zeta\|^2_{H^{\frac{1}{2}}} + \|\delta_\omega\|^2_{H^1(\Omega_t)}\big). \tag{13.8}$$

On the other hand, we know

$$-\Delta(\eta-\eta_1) + (\eta-\eta_1) = \operatorname{div}(\zeta-\zeta_1) = \operatorname{div}\delta_\zeta,$$

which implies

$$\|\eta-\eta_1\|_{H^{\frac{3}{2}}} \leq C\|\delta_\zeta\|_{H^{\frac{1}{2}}}.$$

Then we deduce from (13.6)–(13.8) that

$$\frac{d}{dt}\big(\|\delta_d\|^2_{H^1(\Omega_t)} + \|\delta_\omega\|^2_{H^1(\Omega_t)} + \|\delta_\zeta\|^2_{H^{\frac{1}{2}}}\big) \leq C\big(\|\delta_d\|^2_{H^1(\Omega_t)} + \|\delta_\omega\|^2_{H^1(\Omega_t)} + \|\delta_\zeta\|^2_{H^{\frac{1}{2}}}\big)$$

together with the initial condition

$$\delta_d|_{t=0} = 0, \quad \delta_\omega|_{t=0} = 0, \quad \delta_\zeta|_{t=0} = 0.$$

Gronwall's inequality implies that $\delta_d = 0, \delta_\omega = 0, \delta_\zeta = 0$. This proves (13.4). □

CHAPTER 14

Proof of Theorem 1.1

This chapter is devoted to proving the local well-posedness of the system (1.1)–(1.5) for the low regularity initial data.

14.1. Construction of approximate smooth solution

First of all, we can construct a sequence of smooth functions η_0^n so that

$$\|\eta_0^n - \eta_0\|_{H^{s+\frac{1}{2}}} \longrightarrow 0 \quad \text{as} \quad n \longrightarrow +\infty,$$
$$\eta_0^n(x) + 1 \geq \frac{3}{4}h_0 \quad \text{for } x \in \mathbf{R}^d.$$

Let $\widetilde{v_0}(x,z) = v_0 \circ \Phi(x,z)$, where $\Phi(x,z) = (x, \rho_\delta(x,z))$ with $\rho_\delta(x,z) = z + (1+z)e^{\delta z|D|}\eta_0$. Then we take a sequence of smooth functions $\widetilde{v_0^n}$ so that

$$\|\widetilde{v_0^n} - \widetilde{v_0}\|_{H^{s+\frac{1}{2}}(\mathcal{S})} \longrightarrow 0 \quad \text{as} \quad n \longrightarrow +\infty.$$

Let $v_0^n(x,y) = \widetilde{v_0^n} \circ (\Phi^n)^{-1}(x,y)$, where $\Phi^n(x,z) = (x, \rho_\delta^n(x,z))$ with $\rho_\delta^n(x,z) = z + (1+z)e^{\delta z|D|}\eta_0^n$.

The pressure P_0^n associated with the initial data (v_0^n, η_0^n) is defined by

$$\begin{cases} -\Delta_{x,y} P_0^n = \partial_i (v_0^n)^j \partial_j (v_0^n)^i \quad \text{in} \quad \Omega_0^n, \\ P_0^n|_{y=\eta_0^n(x)} = 0, \quad \partial_y P_0^n|_{y=-1} = -1, \end{cases}$$

where $\Omega_0^n = \{(x,y) : y = \eta_0^n(x), x \in \mathbf{R}^d\}$. Then we can show that

$$\|\nabla_{x,z}(P_0^n \circ \Phi^n - P_0 \circ \Phi)\|_{X^{s-\frac{1}{2}}([-\frac{1}{2},0])} \longrightarrow 0 \quad \text{as} \quad n \longrightarrow +\infty.$$

This ensures that for n big enough, there holds

$$-\partial_y P_0^n\big|_{y=\eta_0^n} \geq \frac{3}{4}c_0.$$

Thus, we apply Theorem 10.1 to obtain a sequence of smooth solutions (v^n, η^n, P^n) associated with the initial data (v_0^n, η_0^n) on a maximal existence time interval $[0, T_n)$.

14.2. Uniform estimates and existence

We denote

$$(V^n, B^n) = v^n|_{y=\eta^n(t,x)}, \quad V_b^n = (v^n)^h|_{y=-1}.$$

We define

$$E_s^n(t) \stackrel{\text{def}}{=} \|(V^n, B^n, V_b^n)(t)\|_{H^s} + \|\eta^n(t)\|_{H^{s+\frac{1}{2}}} + \|\widetilde{\omega^n}(t)\|_{H^{s-\frac{1}{2}}(\mathcal{S})},$$
$$E_{s,l}^n(t) \stackrel{\text{def}}{=} \|(V^n, B^n)(t)\|_{H^{s-\frac{1}{2}}}^2 + \|V_b^n(t)\|_{H^s}^2 + \|\eta^n(t)\|_{H^s}^2 + \|\widetilde{\omega^n}(t)\|_{H^{s-\frac{1}{2}}(\mathcal{S})}.$$

The goal of this subsection is to show that there exits $T > 0$ depending only on $E_s(0)$ and c_0, h_0, s such that that for any $t \in [0, \min(T, T_n))$, there holds

$$E_s^n(t) \le \mathcal{P}(E_s(0)). \tag{14.1}$$

Here and in what follows we denote by $\mathcal{P}$ an increasing function depending only on c_0, h_0, s, which may change from line to line.

First of all, it follows from Proposition 9.3 that

$$\frac{d}{dt}\big(\|(V^n, B^n)\|^2_{H^{s-\frac{1}{2}}} + \|V_b^n\|^2_{H^s} + \|\eta^n\|^2_{H^s}\big) \le \mathcal{P}(E_s^n(t)). \tag{14.2}$$

Next, we estimate the vorticity ω^n, which satisfies

$$\partial_t \widetilde{\omega^n} + \overline{v^n}\cdot\nabla_{x,z}\widetilde{\omega^n} = \widetilde{\omega^n}^h \cdot \big(\nabla\widetilde{v^n} - \frac{\nabla\rho_\delta^n}{\partial_z\rho_\delta^n}\partial_z\widetilde{v^n}\big) + \widetilde{\omega^n}^{d+1}\frac{\partial_z\widetilde{v^n}}{\partial_z\rho_\delta^n} \triangleq F^n,$$

where $\overline{v^n} = \big(\widetilde{v^n}, \frac{1}{\partial_z\rho_\delta^n}(\widetilde{v^n}_{d+1} - \partial_t\rho_\delta^n - \widetilde{v^n}^h\cdot\nabla\rho_\delta^n)\big)$. By Lemma 9.6 and Lemma 2.13, we have

$$\|\overline{v^n}\|_{H^{s+\frac{1}{2}}(\overline{\mathcal{S}})} + \|F^n\|_{H^{s-\frac{1}{2}}(\overline{\mathcal{S}})} \le \mathcal{P}(E_s^n(t)).$$

Let $\overline{v_e^n}$ and F_e^n be the extension of $\overline{v^n}$ and F^n to $\mathbf{R}^{d+1}$ so that

$$\|\overline{v_e^n}\|_{H^{s+\frac{1}{2}}(\mathbf{R}^{d+1})} \le C\|\overline{v^n}\|_{H^{s+\frac{1}{2}}(\overline{\mathcal{S}})},$$
$$\|F_e^n\|_{H^{s-\frac{1}{2}}(\mathbf{R}^{d+1})} \le C\|F^n\|_{H^{s-\frac{1}{2}}(\overline{\mathcal{S}})}.$$

We define $\widetilde{\omega_e^n}$ to be a solution of the following transport equation in $\mathbf{R}^{d+1}$

$$\partial_t\widetilde{\omega_e^n} + \overline{v_e^n}\cdot\nabla_{x,y}\widetilde{\omega_e^n} = F_e^n, \quad \widetilde{\omega_e^n}(0) = \widetilde{\omega_{0,e}^n}(x).$$

It is obvious that $\widetilde{\omega_e^n} = \widetilde{\omega^n}$ in $\overline{\mathcal{S}}$ by the uniqueness of the solution. By a standard $H^{s-\frac{1}{2}}$ energy estimate, we deduce that

$$\frac{d}{dt}\|\widetilde{\omega_e^n}(t)\|_{H^{s-\frac{1}{2}}(\mathbf{R}^{d+1})} \le C\|\overline{v_e^n}\|_{H^{s+\frac{1}{2}}(\mathbf{R}^{d+1})}\|\widetilde{\omega_e^n}(t)\|_{H^{s-\frac{1}{2}}(\mathbf{R}^{d+1})} + \|F_e^n\|_{H^{s-\frac{1}{2}}(\mathbf{R}^{d+1})},$$

from which and Gronwall's inequality, it follows that

$$\|\widetilde{\omega_e^n}(t)\|_{H^{s-\frac{1}{2}}(\mathbf{R}^{d+1})} \le \big(\|\widetilde{\omega_{0,e}^n}\|_{H^{s-\frac{1}{2}}(\mathbf{R}^{d+1})} + \|F_e^n\|_{H^{s-\frac{1}{2}}(\mathbf{R}^{d+1})}\big)e^{\int_0^t \|\overline{v_e^n}\|_{H^{s+\frac{1}{2}}(\mathbf{R}^{d+1})}dt'}.$$

This implies that

$$\|\widetilde{\omega^n}(t)\|_{H^{s-\frac{1}{2}}(\overline{\mathcal{S}})} \le C\big(\|\widetilde{\omega_0^n}\|_{H^{s-\frac{1}{2}}(\overline{\mathcal{S}})} + \mathcal{P}(E_s^n(t))\big)e^{\int_0^t \mathcal{P}(E_s^n(t'))dt'},$$

which together with (14.2) leads to

$$E_{s,l}^n(t) \le E_{s,l}^n(0) + \int_0^t \mathcal{P}(E_s^n(t'))dt'. \tag{14.3}$$

Now let us turn to the higher order energy estimate. For this, we need the following refined elliptic estimate from [**3**].

LEMMA 14.1. *Let $v \in H^{s+\frac{1}{2}}(\overline{\mathcal{S}})$ be a solution of the elliptic equation* (4.11) *with $v(0) = f$. Then it holds that for any $\sigma \in [-\frac{1}{2}, s-1]$,*

$$\|\nabla_{x,z}v\|_{X^\sigma([-\frac{1}{2},0])} \le \mathcal{P}(\|\eta\|_{H^s})\big(\|f\|_{H^{\sigma+1}} + \|F_0\|_{Y^\sigma(I)}\big),$$

if it holds for $\sigma = -\frac{1}{2}$.

LEMMA 14.2. *It holds that*

$$\|U^n\|_{H^s} + \|(V^n, B^n)\|_{H^s} + \|\eta^n\|_{H^{s+\frac{1}{2}}}$$
$$\leq \mathcal{P}(E^n_{s,l})\big(\|D_t U^n\|_{H^{s-\frac{1}{2}}} + \|T_{\sqrt{a^n\lambda^n}} U^n\|_{H^{s-\frac{1}{2}}}\big) + \mathcal{P}(E^n_{s,l}).$$

PROOF. **Step 1.** Estimate for U^n.

Let $\epsilon \in (0, \min(s_d, 1))$ with $s_d = s - \frac{d}{2} - 1$. Applying Lemma 14.1 to P^n, we obtain

$$\|\nabla_{x,z}\widetilde{P^n_1}\|_{X^{s-1}([-\frac{1}{2},0])} \leq \mathcal{P}(E^n_{s,l}(t)), \quad P^n_1 = P^n + y. \tag{14.4}$$

Then we deduce from Proposition 2.6 and (14.4) that

$$\begin{aligned}\|U^n\|_{H^s} \leq & \|T_{(\sqrt{a^n\lambda^n})^{-1}} T_{\sqrt{a^n\lambda^n}} U^n\|_{H^s} + \|(T_{(\sqrt{a^n\lambda^n})^{-1}} T_{(\sqrt{a^n\lambda^n})} - 1)U^n\|_{H^s} \\ \leq & \mathcal{P}(E^n_{s,l})\|T_{\sqrt{a^n\lambda^n}} U^n\|_{H^{s-\frac{1}{2}}} + \mathcal{P}(E^n_{s,l})\|U^n\|_{H^{s-\epsilon}},\end{aligned}$$

which implies that

$$\|U^n\|_{H^s} \leq \mathcal{P}(E^n_{s,l})\|T_{\sqrt{a^n\lambda^n}} U^n\|_{H^s} + \mathcal{P}(E^n_{s,l}). \tag{14.5}$$

Step 2. Estimate for η^n

By the proof of Lemma 9.16, we know that

$$\zeta^n = T_{(a^n)^{-1}}\big(- D_t U^n + h^n_1 + [D_t, T_{\zeta^n}]B^n\big) + (T_{(a^n)^{-1}} T_{a^n} - 1)\zeta^n,$$

which along with Proposition 2.6 and Proposition 2.21 gives

$$\begin{aligned}\|\zeta^n\|_{H^{s-\frac{1}{2}}} \leq \mathcal{P}(E^n_{s,l})\big(&\|D_t U^n\|_{H^{s-\frac{1}{2}}} + \|\zeta^n\|_{H^{s-\frac{1}{2}-\epsilon}} + \|h^n_1\|_{H^{s-\frac{1}{2}}} \\ &+ \|V^n\|_{B^1_{\infty,1}}\|B^n\|_{H^{s-\frac{1}{2}}} + \|D_t\zeta^n\|_{L^\infty}\|B^n\|_{H^{s-\frac{1}{2}}}\big).\end{aligned}$$

Recall that $h^n_1 = (T_{V^n} - V^n)\cdot\nabla V^n - R(a^n, \zeta^n) + T_{\zeta^n}(T_{V^n} - V^n)\cdot\nabla B^n$. Then Lemma 2.10 and (14.4) ensure that

$$\|h^n_1\|_{H^{s-\frac{1}{2}}} \leq \mathcal{P}(E^n_{s,l})\|V^n\|_{W^{1,\infty}}\|(V^n, B^n)\|_{H^{s-\frac{1}{2}}} + \mathcal{P}(E^n_{s,l})\|\zeta^n\|_{H^{s-\frac{1}{2}-\epsilon}}.$$

Recall that $(\partial_t + V^n\cdot\nabla)\zeta^n = \nabla B^n - \nabla V^n\cdot\nabla\eta^n$. Then we have

$$\|D_t\zeta^n(t)\|_{L^\infty} \leq C\|(V^n, B^n)\|_{W^{1,\infty}}\|\eta^n\|_{W^{1,\infty}}. \tag{14.6}$$

Thanks to $s > 1 + \frac{d}{2}$, we get by the interpolation that

$$\|(V^n, B^n)\|_{B^1_{\infty,1}} \leq \|(V^n, B^n)\|^{2\epsilon}_{H^{s-\frac{1}{2}}}\|(V^n, B^n)\|^{1-2\epsilon}_{H^s}. \tag{14.7}$$

Thus, we conclude that

(14.8)
$$\|\eta^n\|_{H^{s+\frac{1}{2}}} \leq \mathcal{P}(E^n_{s,l}(t)) + \mathcal{P}(E^n_{s,l}(t))\|D_t U^n\|_{H^{s-\frac{1}{2}}} + \mathcal{P}(E^n_{s,l}(t))\|V^n, B^n\|^{1-2\epsilon}_{H^s}.$$

Step 3. Estimates for B^n.

By the proof of Lemma 9.17, we know that

$$\mathrm{div} U^n = -T_{q^n}B^n - R(\eta^n)B^n + V^n_\omega + T_{\mathrm{div}\zeta^n}B^n,$$

where the symbol $q^n = \lambda^n - i\zeta^n\cdot\xi$ and

$$V^n_\omega = -\partial_y(v^n)^{d+1}_\omega + \nabla\eta^n\cdot\nabla(v^n)^{d+1}_\omega\big|_{y=\eta^n} + \partial_i\eta^n\omega^n_{d+1,i}\big|_{y=\eta^n}.$$

By using Lemma 14.1 and following the proof of Proposition 5.4, we can deduce that

$$\begin{aligned}&\|R(\eta^n)(V^n,B^n)\|_{H^{s-1}} \le \mathcal{P}(E^n_{s,\ell}(t))\|\eta^n\|_{H^{s+\frac12}},\\ (14.9)\qquad &\|\nabla_{x,z}\widetilde{v^n_\omega}\|_{X^{s-1}([-\frac12,0])} \le \mathcal{P}(E^n_{s,l}(t)).\end{aligned}$$

Thus, we infer from Lemma 2.10 that

$$\begin{aligned}\|T_{q^n}B^n\|_{H^{s-1}} \le&\|U^n\|_{H^s}+\|R(\eta^n)B^n\|_{H^{s-1}}+\|V^n_\omega\|_{H^{s-1}}+\|\zeta^n\|_{C^\epsilon}\|B^n\|_{H^{s-\epsilon}}\\ (14.10)\qquad \le&\|U^n\|_{H^s}+\mathcal{P}(E^n_{s,l})\|\eta^n\|_{H^{s+\frac12}}+\mathcal{P}(E^n_{s,l})\|B^n\|_{H^{s-\epsilon}}.\end{aligned}$$

Then Proposition 2.6 ensures that

$$\begin{aligned}\|B^n\|_{H^s} \le&\|(T_{(q^n)^{-1}}T_{q^n}-1)B^n\|_{H^s}+\|T_{(q^n)^{-1}}T_{q^n}B^n\|_{H^s}\\ \le&\mathcal{P}(E^n_{s,l})\|B^n\|_{H^{s-\epsilon}}+\mathcal{P}(E^n_{s,l})\|T_{q^n}B^n\|_{H^{s-1}}\\ (14.11)\qquad \le&\mathcal{P}(E^n_{s,l})(\|B^n\|_{H^{s-\epsilon}}+\|U^n\|_{H^s}+\mathcal{P}(E^n_{s,l})\|\eta^n\|_{H^{s+\frac12}}).\end{aligned}$$

Step 4. Completion of the estimate
Combining the estimates (14.11), (14.10) and (14.5), we have that

$$\begin{aligned}&\|U^n\|_{H^s}+\|B^n\|_{H^s}+\|\eta^n\|_{H^{s+\frac12}}\\ (14.12)\qquad &\le \mathcal{P}(E^n_{s,l})\big(\|D_tU^n\|_{H^{s-\frac12}}+\|T_{\sqrt{a^n\lambda^n}}U^n\|_{H^{s-\frac12}}\big)+\mathcal{P}(E^n_{s,l}).\end{aligned}$$

This also gives

$$\begin{aligned}\|V^n\|_{H^s} \le&\|U^n\|_{H^s}+\|T_{\zeta^n}B^n\|_{H^s}\\ \le&\mathcal{P}(E^n_{s,l})\big(\|D_tU^n\|_{H^{s-\frac12}}+\|T_{\sqrt{a^n\lambda^n}}U^n\|_{H^{s-\frac12}}\big)+\mathcal{P}(E^n_{s,l}),\end{aligned}$$

which together with (14.5) and (14.11) gives the lemma. □

Proceeding the same way as in section 9.6, we can deduce that

$$\begin{aligned}&\frac{d}{dt}\Big(\|D_tU^n\|^2_{H^{s-\frac12}}+\|T_{\sqrt{a^n\lambda^n}}U^n\|^2_{H^{s-\frac12}}+\langle(f^n_\omega)_{s-1},U^n_s\rangle\\ (14.13)\qquad &\quad+\langle g^n_{s-1/2},h^n_{s-1/2}\rangle+\frac12\langle g^n_{s-1/2},g^n_{s-1/2}\rangle\Big)\le\mathcal{P}(E^n_s).\end{aligned}$$

Here $g^n=[D_t,T_{\zeta^n}]B^n$ and $h^n=D_tU^n-[D_t,T_{\zeta^n}]B^n$.
It follows from Proposition 2.21 that

$$\begin{aligned}&\|g^n\|_{H^{s-\frac12}}\le C\big(\|V^n\|_{B^1_{\infty,1}}+\|D_t\zeta^n\|_{L^\infty}\big)\|B^n\|_{H^{s-\frac12}},\\ &\|h^n\|_{H^{s-\frac12}}\le\|D_tU^n\|_{H^{s-\frac12}}+C\big(\|V^n\|_{B^1_{\infty,1}}+\|D_t\zeta^n\|_{L^\infty}\big)\|B^n\|_{H^{s-\frac12}},\end{aligned}$$

and we have by (14.9) that

$$\|f^n_\omega\|_{H^{s-1}}\le\mathcal{P}(E^n_{s,l}(t)).$$

Plugging the above estimates into (14.13), we obtain

$$\begin{aligned}&\|D_tU^n(t)\|^2_{H^{s-\frac12}}+\|T_{\sqrt{a^n\lambda^n}}U^n(t)\|^2_{H^{s-\frac12}}\\ &\le\mathcal{P}(E_s(0))+\mathcal{P}(E^n_{s,l}(t))\|U^n(t)\|_{H^s}+C\big(\|V^n(t)\|_{B^1_{\infty,1}}+\|D_t\zeta^n(t)\|_{L^\infty}\big)^2\|B^n(t)\|^2_{H^{s-\frac12}}\\ &\quad+C\big(\|V^n(t)\|_{B^1_{\infty,1}}+\|D_t\zeta^n(t)\|_{L^\infty}\big)\|B^n(t)\|_{H^{s-\frac12}}\|D_tU^n\|_{H^{s-\frac12}}+\int_0^t\mathcal{P}(E^n_s(t'))dt',\end{aligned}$$

which together with (14.6), (14.7) and Lemma 14.2 implies that

$$E_s^n(t) \le \mathcal{P}(E_s(0)) + \mathcal{P}(E_{s,l}^n(t)) + \int_0^t \mathcal{P}(E_s^n(t'))dt'.$$

This together with (14.3) ensures that there exits $T > 0$ depending on $E_s(0)$ and c_0, h_0, s such that for any $t \in [0, \min(T, T_n))$, there holds

$$E_s^n(t) \le \mathcal{P}(E_s(0)).$$

With the uniform estimates, the existence of the solution can be deduced by a standard compact argument. Here we omit the details.

14.3. Uniqueness of the solution

Let (η^1, v^1, P^1) and (η^2, v^2, P^2) be two solutions of the system (1.1)–(1.5) with the same initial data. Assume that the solutions satisfy

$$\sup_{t\in[0,T]} \big(\|\eta^1(t)\|_{H^{s+\frac12}} + \|\widetilde{v^1}(t)\|_{H^{s+\frac12}(\overline{\mathcal{S}})} + \|\eta^2(t)\|_{H^{s+\frac12}} + \|\widetilde{v^2}(t)\|_{H^{s+\frac12}(\overline{\mathcal{S}})}\big) \le \mathcal{P}_0,$$

Here and in what follows, we denote by $\mathcal{P}_0$ a constant depending only on $E_s(0), c_0, h_0, s$, which may change from line to line.

We denote $\delta_f = f^2 - f^1$ and introduce

$$\delta_{\mathcal{E}}(t) \overset{\text{def}}{=} \delta_{\mathcal{E}_1}(t) + \delta_{\mathcal{E}_2}(t),$$

where $\delta_{\mathcal{E}_1}(t)$ and $\delta_{\mathcal{E}_2}(t)$ are given by

$$\begin{aligned}\delta_{\mathcal{E}_1}(t) =& \|(\delta_V, \delta_B)(t)\|_{H^{s-1}} + \|\delta_\eta(t)\|_{H^{s-\frac12}},\\ \delta_{\mathcal{E}_2}(t) =& \|(\delta_V, \delta_B)(t)\|_{H^{s-\frac32}} + \|(\delta_{V_b}, \delta_\eta)(t)\|_{H^{s-1}} + \|\delta_{\widetilde{\omega}}(t)\|_{H^{s-\frac32}(\overline{\mathcal{S}})}.\end{aligned}$$

In what follows, we denote by $L_i (i = 0, 1, 2)$ some nonlinear term, which satisfies

$$\|L_i(t)\|_{H^{s-1-\frac{i}{2}}} \le \mathcal{P}_0 \delta_{\mathcal{E}}(t).$$

We denote $D_t = \partial_t + V^2 \cdot \nabla$.

Following the proof of section 12.2, we can show that

(14.14) $\|D_t \delta_{R_\omega}\|_{H^{s-2}} + \|(\delta_{R(\eta)(V,B)}, \delta_a)\|_{H^{s-\frac32}} + \|\delta_{\nabla P|_{y=-1}}\|_{H^{s-1}} \le \mathcal{P}_0 \delta_{\mathcal{E}}(t),$

(14.15) $\|\delta_{R_\omega}\|_{H^{s-2}} \le \mathcal{P}_0 \delta_{\mathcal{E}_2}(t).$

With (14.14) and (14.15), we can deduce that

$$\begin{cases} D_t(\delta_V, \delta_B) = L_1,\\ (\partial_t + V_b^2 \cdot \nabla)\delta_{V_b} = L_0,\\ D_t \delta_\zeta = L_2,\\ D_t \delta_U + T_{a^2\lambda^2}\delta_U = L_1 + T_{a^2}\delta_{R_\omega}.\end{cases}$$

The energy estimate ensures that

$$\begin{aligned}&\|D_t \delta_U(t)\|_{H^{s-\frac32}} + \|T_{\sqrt{a^2\lambda^2}}\delta_U(t)\|_{H^{s-\frac32}} + \|(\delta_V, \delta_B)(t)\|_{H^{s-\frac32}}\\ &+ \|(\delta_{V_b}, \delta_\eta)(t)\|_{H^{s-1}} \le \mathcal{P}_0 \int_0^t \delta_{\mathcal{E}}(t')dt'.\end{aligned}$$

For the vorticity, we have

$$\|\delta_{\widetilde{\omega}}(t)\|_{H^{s-\frac32}(\overline{\mathcal{S}})} \le \mathcal{P}_0 \int_0^t \delta_{\mathcal{E}}(t')dt'.$$

Using the fact that

$$\delta_{\mathcal{E}}(t) \leq \mathcal{P}_0\big(\delta_{\mathcal{E}_2}(t) + \|D_t\delta_U(t)\|_{H^{s-\frac{3}{2}}} + \|T_{\sqrt{a^2\lambda^2}}\delta_U(t)\|_{H^{s-\frac{3}{2}}}\big).$$

We conclude that

$$\delta_{\mathcal{E}}(t) \leq \mathcal{P}_0 \int_0^t \delta_{\mathcal{E}}(t')dt',$$

which implies that $\delta_{\mathcal{E}}(t) = 0$ for any $t \in [0, T]$. The uniqueness is proved. □

Acknowledgement

Zhifei Zhang is partially supported by NSF of China under Grant 11425103.

Bibliography

[1] T. Alazard, N. Burq, and C. Zuily, *On the water-wave equations with surface tension*, Duke Math. J. **158** (2011), no. 3, 413–499, DOI 10.1215/00127094-1345653. MR2805065

[2] T. Alazard, N. Burq, and C. Zuily, *Strichartz estimates for water waves* (English, with English and French summaries), Ann. Sci. Éc. Norm. Supér. (4) **44** (2011), no. 5, 855–903, DOI 10.24033/asens.2156. MR2931520

[3] T. Alazard, N. Burq, and C. Zuily, *On the Cauchy problem for gravity water waves*, Invent. Math. **198** (2014), no. 1, 71–163, DOI 10.1007/s00222-014-0498-z. MR3260858

[4] T. Alazard and G. Métivier, *Paralinearization of the Dirichlet to Neumann operator, and regularity of three-dimensional water waves*, Comm. Partial Differential Equations **34** (2009), no. 10-12, 1632–1704, DOI 10.1080/03605300903296736. MR2581986

[5] T. Alazard, N. Burq, and C. Zuily, *Strichartz estimates and the Cauchy problem for the gravity water waves equations*, Mem. Amer. Math. Soc. **256** (2018), no. 1229, v+108, DOI 10.1090/memo/1229. MR3852259

[6] T. Alazard and J.-M. Delort, *Global solutions and asymptotic behavior for two dimensional gravity water waves* (English, with English and French summaries), Ann. Sci. Éc. Norm. Supér. (4) **48** (2015), no. 5, 1149–1238, DOI 10.24033/asens.2268. MR3429478

[7] D. M. Ambrose and N. Masmoudi, *The zero surface tension limit of two-dimensional water waves*, Comm. Pure Appl. Math. **58** (2005), no. 10, 1287–1315, DOI 10.1002/cpa.20085. MR2162781

[8] D. M. Ambrose and N. Masmoudi, *The zero surface tension limit of three-dimensional water waves*, Indiana Univ. Math. J. **58** (2009), no. 2, 479–521, DOI 10.1512/iumj.2009.58.3450. MR2514378

[9] H. Bahouri, J.-Y. Chemin, and R. Danchin, *Fourier analysis and nonlinear partial differential equations*, Grundlehren der Mathematischen Wissenschaften [Fundamental Principles of Mathematical Sciences], vol. 343, Springer, Heidelberg, 2011. MR2768550

[10] J. T. Beale, T. Kato, and A. Majda, *Remarks on the breakdown of smooth solutions for the 3-D Euler equations*, Comm. Math. Phys. **94** (1984), no. 1, 61–66. MR763762

[11] J. Bourgain and D. Li, *Strong ill-posedness of the incompressible Euler equation in borderline Sobolev spaces*, Invent. Math. **201** (2015), no. 1, 97–157, DOI 10.1007/s00222-014-0548-6. MR3359050

[12] Á. Castro, D. Córdoba, C. Fefferman, F. Gancedo, and M. López-Fernández, *Rayleigh-Taylor breakdown for the Muskat problem with applications to water waves*, Ann. of Math. (2) **175** (2012), no. 2, 909–948, DOI 10.4007/annals.2012.175.2.9. MR2993754

[13] A. Castro, D. Córdoba, C. Fefferman, F. Gancedo, and J. Gómez-Serrano, *Finite time singularities for the free boundary incompressible Euler equations*, Ann. of Math. (2) **178** (2013), no. 3, 1061–1134, DOI 10.4007/annals.2013.178.3.6. MR3092476

[14] J.-Y. Chemin and N. Lerner, *Flot de champs de vecteurs non lipschitziens et équations de Navier-Stokes* (French), J. Differential Equations **121** (1995), no. 2, 314–328, DOI 10.1006/jdeq.1995.1131. MR1354312

[15] D. Christodoulou and H. Lindblad, *On the motion of the free surface of a liquid*, Comm. Pure Appl. Math. **53** (2000), no. 12, 1536–1602, DOI 10.1002/1097-0312(200012)53:12⟨1536::AID-CPA2⟩3.3.CO;2-H. MR1780703

[16] D. Coutand and S. Shkoller, *Well-posedness of the free-surface incompressible Euler equations with or without surface tension*, J. Amer. Math. Soc. **20** (2007), no. 3, 829–930, DOI 10.1090/S0894-0347-07-00556-5. MR2291920

[17] D. Coutand and S. Shkoller, *On the finite-time splash and splat singularities for the 3-D free-surface Euler equations*, Comm. Math. Phys. **325** (2014), no. 1, 143–183, DOI 10.1007/s00220-013-1855-2. MR3147437

[18] W. Craig, *An existence theory for water waves and the Boussinesq and Korteweg-de Vries scaling limits*, Comm. Partial Differential Equations **10** (1985), no. 8, 787–1003, DOI 10.1080/03605308508820396. MR795808

[19] V. Kreĭg and K. E. Veĭn, *Mathematical aspects of surface waves on water* (Russian, with Russian summary), Uspekhi Mat. Nauk **62** (2007), no. 3(375), 95–116, DOI 10.1070/RM2007v062n03ABEH004413; English transl., Russian Math. Surveys **62** (2007), no. 3, 453–473. MR2355420

[20] P. Germain, N. Masmoudi, and J. Shatah, *Global solutions for the gravity water waves equation in dimension 3*, Ann. of Math. (2) **175** (2012), no. 2, 691–754, DOI 10.4007/annals.2012.175.2.6. MR2993751

[21] T. Kato and G. Ponce, *Commutator estimates and the Euler and Navier-Stokes equations*, Comm. Pure Appl. Math. **41** (1988), no. 7, 891–907, DOI 10.1002/cpa.3160410704. MR951744

[22] J. K. Hunter, M. Ifrim, and D. Tataru, *Two dimensional water waves in holomorphic coordinates*, Comm. Math. Phys. **346** (2016), no. 2, 483–552, DOI 10.1007/s00220-016-2708-6. MR3535894

[23] M. Ifrim and D. Tataru, *Two dimensional water waves in holomorphic coordinates II: Global solutions*, Bull. Soc. Math. France **144** (2016), no. 2, 369–394, DOI 10.24033/bsmf.2717. MR3499085

[24] D. Lannes, *Well-posedness of the water-waves equations*, J. Amer. Math. Soc. **18** (2005), no. 3, 605–654, DOI 10.1090/S0894-0347-05-00484-4. MR2138139

[25] G. Métivier, *Para-differential calculus and applications to the Cauchy problem for nonlinear systems*, Centro di Ricerca Matematica Ennio De Giorgi (CRM) Series, vol. 5, Edizioni della Normale, Pisa, 2008. MR2418072

[26] A. D. Ionescu and F. Pusateri, *Global solutions for the gravity water waves system in 2d*, Invent. Math. **199** (2015), no. 3, 653–804, DOI 10.1007/s00222-014-0521-4. MR3314514

[27] H. Lindblad, *Well-posedness for the motion of an incompressible liquid with free surface boundary*, Ann. of Math. (2) **162** (2005), no. 1, 109–194, DOI 10.4007/annals.2005.162.109. MR2178961

[28] M. Ming and Z. Zhang, *Well-posedness of the water-wave problem with surface tension*, J. Math. Pures Appl. (9) **92** (2009), no. 5, 429–455, DOI 10.1016/j.matpur.2009.05.005. MR2558419

[29] V. I. Nalimov, *The Cauchy-Poisson problem* (Russian), Dinamika Splošn. Sredy **Vyp. 18 Dinamika Židkost. so Svobod. Granicami** (1974), 104–210, 254. MR0609882

[30] J. Shatah and C. Zeng, *Geometry and a priori estimates for free boundary problems of the Euler equation*, Comm. Pure Appl. Math. **61** (2008), no. 5, 698–744, DOI 10.1002/cpa.20213. MR2388661

[31] J. Shatah and C. Zeng, *A priori estimates for fluid interface problems*, Comm. Pure Appl. Math. **61** (2008), no. 6, 848–876, DOI 10.1002/cpa.20241. MR2400608

[32] J. Shatah and C. Zeng, *Local well-posedness for fluid interface problems*, Arch. Ration. Mech. Anal. **199** (2011), no. 2, 653–705, DOI 10.1007/s00205-010-0335-5. MR2763036

[33] C. Wang and Z. Zhang, *Break-down criterion for the water-wave equation*, Sci. China Math. **60** (2017), no. 1, 21–58, DOI 10.1007/s11425-016-5141-6. MR3585049

[34] S. Wu, *Well-posedness in Sobolev spaces of the full water wave problem in 2-D*, Invent. Math. **130** (1997), no. 1, 39–72, DOI 10.1007/s002220050177. MR1471885

[35] S. Wu, *Well-posedness in Sobolev spaces of the full water wave problem in 3-D*, J. Amer. Math. Soc. **12** (1999), no. 2, 445–495, DOI 10.1090/S0894-0347-99-00290-8. MR1641609

[36] S. Wu, *Almost global wellposedness of the 2-D full water wave problem*, Invent. Math. **177** (2009), no. 1, 45–135, DOI 10.1007/s00222-009-0176-8. MR2507638

[37] S. Wu, *Global wellposedness of the 3-D full water wave problem*, Invent. Math. **184** (2011), no. 1, 125–220, DOI 10.1007/s00222-010-0288-1. MR2782254

[38] S. Wu, *On a class of self-similar 2D surface water waves*, arXiv:1206.2208.

[39] H. Yosihara, *Gravity waves on the free surface of an incompressible perfect fluid of finite depth*, Publ. Res. Inst. Math. Sci. **18** (1982), no. 1, 49–96, DOI 10.2977/prims/1195184016. MR660822

[40] P. Zhang and Z. Zhang, *On the free boundary problem of three-dimensional incompressible Euler equations*, Comm. Pure Appl. Math. **61** (2008), no. 7, 877–940, DOI 10.1002/cpa.20226. MR2410409

[41] Z.-f. Zhang, *Global well-posedness for the 2D critical dissipative quasi-geostrophic equation*, Sci. China Ser. A **50** (2007), no. 4, 485–494, DOI 10.1007/s11425-007-0002-y. MR2333748

Editorial Information

To be published in the *Memoirs*, a paper must be correct, new, nontrivial, and significant. Further, it must be well written and of interest to a substantial number of mathematicians. Piecemeal results, such as an inconclusive step toward an unproved major theorem or a minor variation on a known result, are in general not acceptable for publication.

Papers appearing in *Memoirs* are generally at least 80 and not more than 200 published pages in length. Papers less than 80 or more than 200 published pages require the approval of the Managing Editor of the Transactions/Memoirs Editorial Board. Published pages are the same size as those generated in the style files provided for $\mathcal{A}_{\mathcal{M}}\mathcal{S}$-LaTeX or $\mathcal{A}_{\mathcal{M}}\mathcal{S}$-TeX.

Information on the backlog for this journal can be found on the AMS website starting from `http://www.ams.org/memo`.

A Consent to Publish is required before we can begin processing your paper. After a paper is accepted for publication, the Providence office will send a Consent to Publish and Copyright Agreement to all authors of the paper. By submitting a paper to the *Memoirs*, authors certify that the results have not been submitted to nor are they under consideration for publication by another journal, conference proceedings, or similar publication.

Information for Authors

Memoirs is an author-prepared publication. Once formatted for print and on-line publication, articles will be published as is with the addition of AMS-prepared frontmatter and backmatter. Articles are not copyedited; however, confirmation copy will be sent to the authors.

Initial submission. The AMS uses Centralized Manuscript Processing for initial submissions. Authors should submit a PDF file using the Initial Manuscript Submission form found at `www.ams.org/submission/memo`, or send one copy of the manuscript to the following address: Centralized Manuscript Processing, MEMOIRS OF THE AMS, 201 Charles Street, Providence, RI 02904-2294 USA. If a paper copy is being forwarded to the AMS, indicate that it is for *Memoirs* and include the name of the corresponding author, contact information such as email address or mailing address, and the name of an appropriate Editor to review the paper (see the list of Editors below).

The paper must contain a *descriptive title* and an *abstract* that summarizes the article in language suitable for workers in the general field (algebra, analysis, etc.). The *descriptive title* should be short, but informative; useless or vague phrases such as "some remarks about" or "concerning" should be avoided. The *abstract* should be at least one complete sentence, and at most 300 words. Included with the footnotes to the paper should be the 2010 *Mathematics Subject Classification* representing the primary and secondary subjects of the article. The classifications are accessible from `www.ams.org/msc/`. The Mathematics Subject Classification footnote may be followed by a list of *key words and phrases* describing the subject matter of the article and taken from it. Journal abbreviations used in bibliographies are listed in the latest *Mathematical Reviews* annual index. The series abbreviations are also accessible from `www.ams.org/msnhtml/serials.pdf`. To help in preparing and verifying references, the AMS offers MR Lookup, a Reference Tool for Linking, at `www.ams.org/mrlookup/`.

Electronically prepared manuscripts. The AMS encourages electronically prepared manuscripts, with a strong preference for $\mathcal{A}_{\mathcal{M}}\mathcal{S}$-LaTeX. To this end, the Society has prepared $\mathcal{A}_{\mathcal{M}}\mathcal{S}$-LaTeX author packages for each AMS publication. Author packages include instructions for preparing electronic manuscripts, samples, and a style file that generates the particular design specifications of that publication series. Though $\mathcal{A}_{\mathcal{M}}\mathcal{S}$-LaTeX is the highly preferred format of TeX, author packages are also available in $\mathcal{A}_{\mathcal{M}}\mathcal{S}$-TeX.

Authors may retrieve an author package for *Memoirs of the AMS* from `www.ams.org/journals/memo/memoauthorpac.html`. The *AMS Author Handbook* is available in PDF format from the author package link. The author package can also be obtained free of charge by sending email to **tech-support@ams.org** or from the Publication Division, American Mathematical Society, 201 Charles St., Providence, RI 02904-2294, USA. When requesting an author package, please specify $\mathcal{A}\mathcal{M}\mathcal{S}$-LaTeX or $\mathcal{A}\mathcal{M}\mathcal{S}$-TeX and the publication in which your paper will appear. Please be sure to include your complete mailing address.

After acceptance. The source files for the final version of the electronic manuscript should be sent to the Providence office immediately after the paper has been accepted for publication. The author should also submit a PDF of the final version of the paper to the editor, who will forward a copy to the Providence office.

Accepted electronically prepared files can be submitted via the web at www.ams.org/submit-book-journal/, sent via FTP, or sent on CD to the Electronic Prepress Department, American Mathematical Society, 201 Charles Street, Providence, RI 02904-2294 USA. TeX source files and graphic files can be transferred over the Internet by FTP to the Internet node ftp.ams.org (130.44.1.100). When sending a manuscript electronically via CD, please be sure to include a message indicating that the paper is for the *Memoirs*.

Electronic graphics. Comprehensive instructions on preparing graphics are available at www.ams.org/authors/journals.html. A few of the major requirements are given here.

Submit files for graphics as EPS (Encapsulated PostScript) files. This includes graphics originated via a graphics application as well as scanned photographs or other computer-generated images. If this is not possible, TIFF files are acceptable as long as they can be opened in Adobe Photoshop or Illustrator.

Authors using graphics packages for the creation of electronic art should also avoid the use of any lines thinner than 0.5 points in width. Many graphics packages allow the user to specify a “hairline” for a very thin line. Hairlines often look acceptable when proofed on a typical laser printer. However, when produced on a high-resolution laser imagesetter, hairlines become nearly invisible and will be lost entirely in the final printing process.

Screens should be set to values between 15% and 85%. Screens which fall outside of this range are too light or too dark to print correctly. Variations of screens within a graphic should be no less than 10%.

Any graphics created in color will be rendered in grayscale for the printed version unless color printing is authorized by the Managing Editor and the Publisher. In general, color graphics will appear in color in the online version.

Inquiries. Any inquiries concerning a paper that has been accepted for publication should be sent to memo-query@ams.org or directly to the Electronic Prepress Department, American Mathematical Society, 201 Charles St., Providence, RI 02904-2294 USA.

Editors

SELECTED PUBLISHED TITLES IN THIS SERIES

For a complete list of titles in this series, visit the
AMS Bookstore at **www.ams.org/bookstore/memoseries/**.